35.00

RY

PATTERNS OF SOCIAL AND TECHNOLOGICAL CHANGE IN EUROPE

To our children
Simon Hochgerner
Rosa Campbell
Andreas Alexandros Katsikides

for a peaceful Europe....

Patterns of Social and Technological Change in Europe

Edited by
SAVVAS KATSIKIDES
MIKE CAMPBELL
JOSEF HOCHGERNER

Avebury
Aldershot · Brookfield USA · Hong Kong · Singapore · Sydney

Published by
Avebury
Ashgate Publishing Limited
Gower House
Croft Road
Aldershot
Hants GU11 3HR
England

Ashgate Publishing Company
Old Post Road
Brookfield
Vermont 05036
USA

British Library Cataloguing in Publication Data

Patterns of Social and Technological
Change in Europe
I. Savvas Katsikides
303.48094

ISBN 1 85628 608 8

Printed and Bound in Great Britain by
Athenaeum Press Ltd, Newcastle upon Tyne.

Contents

Figures and tables

List of contributors

Reinhard Bachmann, University of Dortmund, Germany

Gerhard Berka, Centre for Social Innovation, Vienna, Austria

Eva Buchinger, Research Centre, Seibersdorf, Austria

Mike Campbell, Leeds Metropolitan University, UK

Steve Cockerill, Leeds Metropolitan University, UK

John Douglas, Leeds Metropolitan University, UK

Ulrike Felt, University of Vienna, Austria

Adolf Filácek, Academy of Science, Czech Republic

Peter Fleissner, University of Technology, Vienna, Austria

Eva Frydová, Academy of Science, Czech Republic

Ruppert W. Gaspar, University of Vienna, Austria

L. Hamilton, Leeds Metropolitan University, UK

H. Hirsch-Kreinsen, University of Darmstadt, Germany

Josef Hochgerner, Centre for Social Innovation, Vienna, Austria

Wolfgang Hofkirchner, University of Technology, Vienna, Austria

Detlef Jahn, University of Paderborn, Germany

Dieter Jaufmann, INIFES, Augsburg, Germany

Savvas Katsikides, University of Technology, Vienna, Austria

Jirí Loudin, Academy of Science, Czech Republic

Peter Mettler, Wiesbaden Polytechnic, Germany

Gerd Möll, University of Dortmund, Germany

Andreas Moschonas, University of Crete, Greece

Zissis Papadimitriou, Aristotelian University of Thessaloniki, Greece

Margit Pohl, University of Technology, Vienna, Austria

Karin Schneider, Siemens, Vienna, Austria

Gerry Stewart, Leeds Metropolitan University, UK

A. Yossifow, Academy of Science, Bulgaria

Acknowledgements

We would like to thank all those who were involved in the publication of this book. In particular, we would like to express special thanks to our publisher Josephine Gooderham for her generous understanding of time limits and her continued support. We owe thanks to our colleague Steve Lowe who unified, set out, proof read and made the final manuscript fit for publishing.

Introduction

The book "Patterns of Social and Technological Changes in Europe" is both informative as well as of analytical value concerning vital issues of European affairs. The basic idea is to draw together research which is devoted to the key questions in the relationship between the various and widely discussed new developments of technological systems and the social changes with which they are associated.

Major social and technological changes are reshaping everyday life all over Europe. Just as there are relevant differences between regions there are also various conceptions of how to study and understand what usually is called the importance of technical progress. Recent developments in particular in the political and economic systems in eastern Europe propel new ideas and increase demands to study the inter-relations between social and technological processes. In particular, authors from all parts of Europe are covering questions such as follows:

Social and cultural factors relating to technical progress, the role and prospects of social research, developments concerning specific professions linked to technical changes, technological standards compared with social structures, cultural identities and economic performance, female and male conceptions of technology, labour relations, developments in the labour market, social limits to technical progress and social impacts thereof, theoretical conceptions and sociological paradigms furthering analysis and understanding of structures and trends in "technological socictics", innovative strategies in technology transfer and research and development issues.

Papers based on research on these fields were written and presented at the "First European Conference of Sociology" at the University of Vienna, Austria, by members of the European research group on ´Patterns of Social and Technological Change in Europe` August 1992. This meeting was organised by the committee of sociology of technology within the Austrian Association of Sociology. The papers were reviewed and prepared for publication by employing the means of the Centre for Social Innovation (CSI) in Vienna.

Mike Campbell Leeds	Josef Hochgerner Vienna	Savvas Katsikides Vienna

Transformation of Patterns I
WORK AND LABOUR RELATIONS

1 The decline of the scientific and technological potential in the former USSR. Some considerations including an 'AI-based' proposal

Reinhard Bachmann and Gerd Möll

Recently it has been commonplace to say that all the states of Central and Eastern Europe which are currently involved in some kind of capitalist transformation are experiencing a rapid decline in economic performance and in standards of living. In some countries the situation has reached the proportions of a severe crisis and the forecasts are anything but optimistic (Marer 1991, OECD 1992). Unemployment, inflation, poverty etc. are characterising the transition to the "free market". The euphoria of liberation and the belief in quickly reaching better living conditions are being replaced by pessimism and deep uncertainty. Nobody seems to have a coherent concept for a successful transition from a centrally-planned to a market-based system. The economic decline is accompanied by moral depression in major parts of the population in the former Communist States, and there is a serious danger that psychological barriers will paralyse the energies of the people which are needed to overcome the present economic and social disaster.

In order to gain some insight into the radical economic and social change, this article will focus on one fundamental socio-economic factor: the production of scientific and technological knowledge. The field of technical progress and innovation is of great importance not only for the economic reconstruction but also for the building of a democratic society. In the following paper we will focus on the case of the (former) USSR and present some theses about (1) the former social and economic structures of knowledge production in the centrally planned economics, and (2) how they are being affected by current changes. We will (3) sketch a theoretical framework for analysing the critical situation and (4) explore a strategy to counter some of the most destructive trends.

1

The scientific and technological potential of the USSR was immense. In 1987 approximately 1.5 million scientific workers were engaged in R&D projects in the Soviet Union (Fortescue 1990). Considering the Soviet Union and Eastern Europe as a whole, it accounted for one third of the world's Ph.D.-level engineers and scientists. Seen from an international perspective, the USSR spent very high rates of its publicly available money on research and development (R&D). However, the Government frequently made complaints about the return on these investments. In order to dissolve this antagonism, it is necessary to look at the specific forms of organising and planning the R&D activities in the Soviet system.

According to Fortescue (1985) we can distinguish between three main characteristics of the R&D system in the Soviet Union: (a) a high degree of formal central planning, counteracted by a substantial degree of de facto decentralisation, (b) the predominance of institution-based planning, and (c) an appreciable lack of co-ordination with production plans.

(a) Under conditions of planned economies innovation strategies and R&D activities organised centrally by the political administrative system. "The general functioning of the R&D sphere is (...) centrally planned and its activities are powerfully influenced by the central administration. R&D units are frequently separated from enterprises and are given research tasks and funding directly from central authorities" (Glugiewicz/Gruchman 1988, 225). On the other hand, this formal structure of centralisation was accompanied by a strong element of de facto decentralisation. Due to a tremendous lack of expertise in high-level political administrative agencies, individual researchers played an important role in deciding what goes into the R&D plans. The research scientists actually had an appreciable degree of autonomy and in this respect had a good chance to get their "pet"-projects accepted. It was no wonder, therefore, that the production of technological knowledge was far too often not oriented towards short-term commercial ends but rather towards long-term basic research problems (Kelly/Salaman 1992).

One problem within this organising innovation mode for the central authorities was to define research priorities. Furthermore, there was no instrument for evaluating whether the results of the selected projects would be of any use for practical applications. "Being dependent on government subsidies (the R&D units) can act without any special risk. One result of this situation is that R&D activities are only weakly linked with the real demand of enterprises for innovations" (Glugiewicz/Gruchman 1988, 225).

(b) Another feature of the Soviet R&D system was the institution-based planning conception. Plans generally were drawn up with regard to a ministry, the "Academy of Sciences" and every big institute and small laboratory, thus there was hardly any chance to get money for a distinguished project on a specific research problem. This mode of planning contributed to inflexibility and a lack of response to new scientific demands. Under these conditions it was not easy to divert attention to new and promising areas of research (Fortescue 1990). A further problem was the over-specialisation of institutes and laboratories and a lack of cross-fertilisation between the scientific disciplines (Fortescue 1985).

(c) In the Soviet system of science planning there was only a very loose co-ordination of R&D with the sphere of production planning. This separation between the sphere of science and the sphere of production made itself felt in the process of implementation. There were "notorious problems in the application of innovative scientific achievements when it comes to realising these on the shop floor level" (Schröder 1988, 434). All too often new products or new processes were handed over to users in an unusable state.

This kind of problem is not unknown to capitalist economies. "Both in planned and market economies the process of innovation tends to become a professional activity and workers and consumers tend to become passive beneficiaries or victims in relation to new technology, rather than subjects taking an active part in the process of innovation" (Lundvall 1988, 365). In the Soviet Union, however, this state of affairs, taken as a whole, seemed to be no reason for great lamentation at the user level, because enterprises had only weak motivations for technological changes in their current activities. In centrally planned economies, "innovations which considerably diminish costs or improve the quality of products are not profitable for enterprises. (...) Innovative activities disturb plan execution, because, regardless of the circumstances, every product, even those which are expensive and of low quality, can be sold" (Glugiewicz/Gruchman 1988, 226). Thus the final producer had no great desire for new technologies. Consequently it was also not surprising that the researchers' motivation was far from being high.

In the late 60's steps were made to reform the planning system by taking into account the so-called "programme-goal method". One important intention of this approach was to alter the planning system into the direction of greater centralisation by strengthening the impact of political decision making. Other key elements of this method were "a stress on the long-term nature of planning and concentration on the planning and funding of specific goal-oriented projects rather than institutions" (Fortescue 1990, 136). Apparently this new approach was not successful in avoiding the pitfalls of the former system. "It seems unlikely that any of the changes of the 1970s or early 1980s significantly changed the essential nature of the traditional system" (Fortescue 1990, 138).

It is important to note that in some respect the situation in the military sector in the Soviet Union was much better than it was in the civil sector. That was not just a matter of having a higher priority when it comes to allocation of resources. In military R&D there was a greater degree of integration between research, development, and production, because the end-user - the military agencies - played an important role throughout the whole science-production cycle. In that way the problem of knowledge transformation was solved considerably better than in the civilian sector. There are also indications that the military R&D had partly solved the problem of knowledge transfer as well by using an effective scientific-technical information service and by means of some specific organisational arrangements (Fortescue 1990). Last but not least, the military R&D system had escaped the motivational problems of the civilian sector by giving scientists better working conditions and the feeling of doing a job based on real social demands. These factors led to a quality of Soviet R&D that in many fields was not too far behind world standards, or even better.

2

To oversimplify somewhat, such was the situation in the field of Soviet R&D when Gorbachev and his reformist followers decided to quit planning economic affairs centrally. From that time on, money was no longer a pure means of economic control functioning as a link between the social needs and the production facilities. As well as this, the government's guarantee of satisfying the elementary needs of the people was taken back. Enterprises were no longer parts of the business cycle of the national economy but institutions that were to make profits. In short: the new agenda called for a radical change in all the forms, means and aims of economic production.

Consequently, in the course of restructuring the Soviet economy, we saw some kind of entrepreneurial revolution in Soviet R&D. "As in so many other areas, Gorbachev took a radical approach to reforms of the science planning system, although without adopting any entirely new concepts or approaches. The basis of this approach is 'full khozraschet (cost-accounting) and self-financing', and as such is intended to be identical with new decentralising procedures introduced into production" (Fortescue 1990, 138). For scientific organisations that meant, that they were compelled to work on a profit-making basis. Due to the fact that public funds were reduced, the R&D institutions were forced to enter arbitrarily into contracts with private production enterprises and other research institutes, which were interested in their products.

Despite all the proclamations and all the plans for a transition from a command to a market-based system, there are serious doubts whether the states of the former USSR are actually moving toward a Western-style capitalism. Their economic systems, some commentators assert, are driven by the pursuit of profit that comes primarily from trade and speculation rather than from transforming production (Burawoy/Krotov 1992). Seen from this point of view, it could be said that they are on the way to "merchant capitalism".

This is a clear indication that the Soviet enterprises, which have been integrated parts of a planned economy and the corresponding division of labour, are not appropriate to be transplanted into a system of capitalist competition. They were not designed for a mode of production, in which the main goal is making profits. The enterprises rather were responsible for ensuring a plan, i.e. the production of a certain amount and kind of commodities. Neither raising the productivity of capital, nor the ceaseless quest for efficiency, was a criterion for production. In the end it was not surprising that changing the raison d'être of production without changing the means of production as well as the consciousness of the working people led to serious trouble.

Following Gorbachev's retirement, the new government in Moscow introduced "free prices" in order to stimulate production. As a result, however, only the product prices were raised enormously. What really happened was just a redistribution of money from buyers to sellers a situation which ultimately led to a disruption in the balanced exchange of production and consumption. While the people were falling back on their last reserves to get their foodstuff, the producing enterprises were left wondering why less and less customers were buying their goods. "Russian factory directors found that few people wanted to buy their goods - especially at free-market prices. Faced with rising costs for raw materials they drastically cut production (namely by 19% in the first 11 months of 1992 from the period in the previous year). Instead of going bankrupt as a western company would in such circumstances, factory managers borrowed - from banks, from the government, sometimes form one another - to pay increasingly idle workers" (Fortune, January 25, 1993).

Thus, introducing "free prices" led to a further decline in production and, in many cases, to a complete breakdown of supplier chains. In order to survive several enterprises built a form of illusionary payment transactions which reminds one of Potemkin's villages. Many firms began to exchange their products in a way which meant a return to a traditional barter economy. Others sold their stockpiles, means of production, real estates, and services for ridiculously cheap prices. Once these phenomena have been considered, one can hardly say that we are witnessing a controlled transition to capitalism. It would be more accurate to say that we are incredulously watching how an economy is gradually destroying its own foundations.

What are the consequences of this for the R&D system of the former USSR? Our main argument states that there are clear signs of a rapid decline of the R&D system of the former USSR. Old structures have been destroyed without being replaced by new ones. Only a very small number of R&D institutes may have a chance to survive under these conditions. Most of the R&D units however, face serious problems or have already been closed. There are just a few newcomers in the field of technological innovation, mostly small firms owned by scientists, who decided to continue research work on their own. The great majority of the R&D workers meanwhile are facing an uncertain future.

In the following section, we will give some explanations of the present situation. For this purpose we will refer to the concepts of knowledge transformation and knowledge transfer, which in some key ideas - especially those concerning the transformation of knowledge - were originally formulated by Malsch (1984). Empirically we have explored these concepts in our recent research work on the development and application of so called "knowledge based systems" within the socio-organisational context of industrial production firms (Bachmann, Moll 1992; Malsch, Bachmann et al.1993). Although the scope of these investigations seems to be remote from our current theme the theoretical concepts of "knowledge transformation" (a) and "knowledge transfer" (b) can also serve as a framework for analysing the transition crisis of Eastern Europe's socio-economical systems and help to generate some further leading theses on how it might be possible to enter the critical situation.

3

(a) One of the central problems which has occurred in the course of the decline in the scientific-technological potential of the former Soviet Union is nearly a complete breakdown of those societal knowledge transformation processes which are a precondition of the development of new and innovative technology. These processes are based on the functionality and coherence of three different stages: knowledge acquisition (a)i, knowledge objectivation (a)ii, and knowledge reintegration (a)iii.

(a)i Knowledge acquisition refers to the generation of basic technological knowledge. Usually, this was the task of big bureaucratic research institutes and laboratories. As these institutions are not functional anymore, the acquisition of innovative basic technological know-how is set out of work. The science system is no longer a stable and relatively well-funded part of the socio-economic system. Now that R&D units, just like any other kind of enterprises, are forced to make profits they have become dependent upon the short-time demands of the sluggishly producing industrial companies. But most of the production firms have no money at all for innovation and research contracts. As a result many R&D institutions are struggling financially. If they cannot earn enough money they are forced to go into liquidation. And that is everyday reality because there is only a slight chance that R&D will find customers in the present situation. Hence, the traditional R&D system comprising many specialised institutes will disappear in time, if no intervention is be made.

For research scientists that means a dramatic decline of opportunities for productive and well-paid employment. Under these circumstances there is not only a danger of systematic "brain drain" from the states of the former USSR. Some Western firms see the eastern R&D potential as a technological gold mine and have already begun to take the pick of the bunch. For example, "Microsoft and Bell Laboratories have hired top Russian scientists and programmers for wages that would make a Third World sweatshop baron blush" (Fortune, January 25, 1993). This enticing away of researchers and scientists by international corporations does not necessarily mean leaving their country to these people. Sometimes the Western firms are not only buying the working power but the rights of using the existing laboratories too. The expenditure for an eastern scientist including laboratory devices is only one tenth of the costs which must be paid in the West (Der Spiegel 12/1990). Due to collapsing structural conditions and to the retreat of highly qualified staff, the formerly high standard of Soviet basic research is allowed to go to waste.

The diminishing employment opportunities are accompanied by a rapid deterioration of working conditions and payment levels. Despite the lack of solid empirical information, it must be assumed that a growing number of research workers is employed under value or unemployed. Following on from this, a considerable loss of morale and motivation amongst the workers is spreading.

(a)ii Knowledge objectivation refers to transposing the results of basic and applied research into technical artefacts. For this purpose on the one hand well trained engineers are needed, who have a high level of experiential skills, qualification, and motivation. On the other hand there has to be a socio-economic system which delivers organisational patterns for conducting development-projects. It might be that the formerly existing structures of developing technological products have been heavy and anything but frictionless. Currently this stage of knowledge transformation seems to be blocked. The old bureaucratic system of planned technological production does not exist any longer; new forms of market pulled motivation have not yet been established. This by necessity has an effect on the system of qualifying young engineers. Under current conditions it seems to be most unreasonable for a young man or woman to invest in a solid engineer's qualification as no one can tell when and how one can make use of this knowledge.

(a)iii By knowledge reintegration we mean appropriate forms of distributing and implementing technology into application contexts. Even if technological systems like production machinery are still completed by engineers, there are no mechanisms to guarantee that they are used reasonably, observing the aim of public welfare. There has been a more or less functional political system of reintegrating technologically objectified knowledge. As the institution of a free and transparent market is not reachable in the current situation there is no institutional property which could be used in order to control the distribution and implementation of technological equipment. It is a consequence of this that the application of technologically "objectivated" knowledge is not obtained satisfactorily. And this is even more problematic as most advanced technological systems - like computers - are not only needed for the consumption industries but also in the R&D-laboratories itself the following cycle of knowledge transformation is blocked from the beginning as some fields of basic research are constitutively based on the existence of advanced technological systems.

Knowledge transformative processes are not only jeopardised by blockages at one of the three stages. Most important is the functioning of the "transmission belt", which links the stages. In the Western states there are constant debates on how to link up the stages in a way that guarantees a minimum of knowledge loss. The political research institutions for example are most interested in how to transpose results from basic research to the production of commercially usable technologies, which shows that there are necessities for central political support and control within the transformative knowledge cycle. In addition to this, the Japanese Government is aware of its political function, as the developing process of new technologies has to overcome obstacles, for which market pulls alone are not strong enough (Freeman 1988). It is a severe misunderstanding of modern capitalist economies if the market - as the new established governments tend to do - is seen as the only propelling power of innovation.

(b) On the level of advanced scientific-technological innovation, the organising of knowledge transfers is another unalterable precondition of success. This means that there have to be structures of communication between the different places of technological R&D. In the former Soviet Union the military in particular kept an eye on this problem and seemed to fulfil this function of systematic knowledge transfers. The "military pull" of the past was obviously strong enough to organise a constant knowledge flow between the most important research institutes. As the military has lost its function as one of the most important institutional complexes in society, there is no one who cares about the linkage of separated knowledge islands. Another problem is that former working contacts between the R&D units are cut due to the decomposition of the Soviet Imperium into interdependent states. "The excessive degree of specialisation which characterised research efforts in the USSR means that individual R&D laboratories may not be able to operate independently now that they belong to different Republics" (Kelly/Salaman 1992). A similar effect arises as a result of the growing economic competition between the R&D units. This just reinforces the trend of the atomisation of scientific-technological knowledge.

Knowledge transfers are not only important on the level of basic research, but also on the level of knowledge objectivation as well as on the level of knowledge reintegration. In the western states mechanisms have been established which allow one to communicate experiences on all these levels. Despite the fact of competition between private firms they have a strong interest to cooperate on one or another level (Bieber/Möll 1993). Only one example is the close co-operation of the big German car producing companies in the field of cost intensive AI research in recent years.

As there is no political will to re-establish the old structures of the USSR it will be one of the major political tasks of the next decades to guide the way out of crisis by offering new socio-economic structures, which could resemble those in the Western European states. The idea that one just has to remove the economic system which has been imposed by the communist authorities in order to receive common welfare has turned out to be a fairy tale. At present, nothing is more important than political visions and programmes.

The states of the former USSR depend on a legacy which is most precious in the present situation - the already atomised and paralysed, yet still existing, scientific-technological know how. It could serve as an idea, which helps to produce solidarity in society and identification for its members. It is mere ideology to believe that a society can change everything overnight. In order to give the change a chance continuities are needed. If the military legacy is neglected, the self-consciousness of having an appreciable part of the world's scientific and technological knowledge could serve as a key to a societal vision of the future.

There are not many possibilities to bridge the past with a totally changed future. Science and technology have been major features of the communist dream and it is also a precondition of the desired capitalist wealth. It is the most precious outcome of 70 years of communism. It would be a fatal political fault not to recognise the chance. But this means to interfere politically and not just to remove restrictions and watch what will happen. What is needed are intelligent forms of knowledge management. In this context it will be necessary to impose new structures which enable people to act, for societal structures are not only restrictive to agency, but also enabling it (Giddens 1984).

4

In as far as knowledge management is concerned, we have learned a lot during the euphoric period of AI in the 1980s. In particular in the field of "expert systems", which has been built with the aim to support - sometimes to replace - human experts in industrial production, medicine and other areas of application. The systems themselves too often ended up as software ruins. But there has been on side effect which is most valuable: We now know more about the management of expert knowledge and the conditions and consequences of its transformation and transfer than we did before. In particular we have learned that expert knowledge is anything but clear cut and stored in the brains of individual holders. Rather it is an integrated system of collective practice. In the following we will explain this and draw some conclusions with regard to the management of R&D knowledge and the preservation of the scientific and technological potential of the former USSR.

(a) Expert knowledge, such as R&D-knowledge, does not exist just in the minds of lonesome individuals. In the context of expert system developing projects the "extractionist" approach (Smith 1985; Waterman 1986), which represents the assumption, that the knowledge engineer's task is nothing but "extracting" a single competent expert's knowledge out of his/her brain, has been recognised as deficient. Practical experiences in knowledge acquisition have made it quite clear that the expert's brain is not something like a case which contains all the knowledge which is needed. However it is more likely to be something resembling a co-ordination centre which is permanently dependent on an external knowledge supply. Thus, the knowledge engineer has to discard the assumption that he/she could simply acquire a single individual's expertise. More and more co-operating experts emerge and thus a network of knowledge transferring channels appears. If parts of this are ignored or cut away, the rest loses its functionality.

Empirically analysing knowledge acquisition reveals the fact that expert knowledge is basically distributed knowledge which makes an innovative organisation function. Concerning a socio-economic system we would claim the same. It is unrealistic to regard such systems as a patchwork of separated knowledge fragments carried each by an individual expert. Rather it is a complex system of interdependencies.

(b) Expert knowledge can a limine not be purified from heuristic strategies and "idiosyncratic" intuitions, motivation, and other non-propositional elements of knowledge. Experiences in knowledge engineering have shown that knowledge acquisition means to establish a dialogue between the experts and the knowledge engineer which firstly means to find a common level of communication and secondly to enter a situation of mutual learning, as there is a very important dimension of mutual acknowledgement and motivation (Mill 1992), which should not be ignored as it is an unalterable part of expertise itself as it seems. If it is missing expertise is nothing than a dead body of rules which are useless in regard to solving real world problems.

Expert knowledge is also constitutively bound to social interests and cultural patterns of meaning. That means that one also has to acknowledge the fact that expertise loses much of its value if the experts have no orientation to a societal system of social needs and individually interesting incentives. The same problem occurs if the cultural origins of expertise is cut away. This, by the way, is one of the reasons which leads western companies to the assumption it could better reap the rewards of using eastern capacities in their original environment than to take the experts to their western research bases.

(c) Expert knowledge does not only have the form of actors' consciousness. Building expert systems means to transform knowledge which is bound to human brains into a kind of knowledge which can be used without any respect of these restrictions. But expertise is not only materialised in technological systems which are called "knowledge based", but in all kinds of technological artefacts (Orlikowski 1992). Technologically "objectivated" knowledge can be developed in apt application contexts, which are distinguished by certain forms of brain-bound qualification and appropriate forms of knowledge which is incorporated in social institutions (Berger, Luckmann 1966), and organisational structures of social reality.

In the case of our subject we can conclude that expert knowledge should not be conserved solely in its brain-bound form as the maintenance of knowledge also demands to care about other aggregates of knowledge which have to be linkable to the actor's consciousness. Therefore it is necessary to maintain a suitable social and organisational infrastructure in order to keep up the scientific-technological potential.

(d) Expert knowledge can not be conserved without using it. One of most distinguishable characteristics of expert knowledge seems to be that it very quickly becomes desolate if one does not use it. Our experiences in the field of AI clearly show what happens if one relies on frozen stocks of knowledge. In particular those cases in which "successful" expert systems support human expertise to a high degree the users tend to get negligent of the renewal of their qualification and have to suffer effects of disqualification. Thus in the end it is not seldom that seemingly "successful" developing projects turn out to be an "own goal", in particular as also the maintenance of these systems requires solid domain knowledge (Sviokla 1986).

From this we conclude that it would be no helpful strategy to store away the scientific-technological know-how of the former Soviet Union and wait for better times. Advanced knowledge is a permanent transition and needs to be embedded in a social -economic context apt for its (re-)generation and application. It has to be viewed as an integral part of the autochthonous social-economic system which despite all the defects, which might be unalterable presently, has to be included into the analysis and the elaboration of political strategies concerning the preservation of the still existing parts of the scientific and technological know-how of the former USSR.

Regaining control of the R&D potential by the political administration of the "Community of Independent States" is easier to proclaim than to transpose. We are aware of the fact that this would require the investment of huge sums of money which, at the same time, would not be available for consumption purposes. And it also requires a high level of societal solidarity which can generate intrinsic motivation to work for a common future. And finally political power is required and the insight that exercising power does not only restrict actors who are affected. Power does also enable actors to attain their goals (Giddens 1976).

If the states of the former Soviet Union can be prevented to step down to the level of a mere supplier to the Western economic welfare it will be necessary to act quickly and try to save some core parts of the rapidly vanishing scientific and technological potential of these states. In order to do so several preconditions have to be fulfilled, which we have not discussed at length in this text. Instead we have focused on two major aspects of knowledge management and innovation, i.e. the transformation and the transfer of scientific-technological knowledge and also gained some insights into the specific character of expert knowledge from the field of Artificial Intelligence.

In order to stop the further decline of the technological potential of the former USSR, it will not be sufficient to try to keep some excellent scientists as their knowledge will be useless, if it is not part and parcel of a complex R&D network. The practical essentials of our theoretical considerations in regard to the retrieval of the R&D potential of the former USSR can be summarised as follows:

1 The laboratories and research institutes in which basic research takes place must be conserved and should not set under pressure of market imperatives.

2 The production of advanced technological artefacts can only be guaranteed if there is a functional system of educating and instructing the after-growth of competent engineers.

3 Concerning the reintegration of scientific-technological knowledge, it is necessary to establish a political and technical infrastructure which guarantees that advanced technology very quickly is distributed to apt user organisations.

4 In regard to the reimprovement of the knowledge transfer the political system itself has to move into the role which the military has formerly played. The government therefore has to represent a common societal interest and not combat the Western states militarily, but to compete scientifically.

The political system has to organise, control, and propel the management of innovation because there are no grown structures which could function autonomously. If the government refuses central control, misunderstanding it as a major feature of the old communist system the final breakdown is predictable. Central control is not valuable in itself but the appropriate remedy under current circumstances. If there is no grown structural basement one has to seek "technical" solutions, which might imply a good deal of bureaucracy. Although no one would defend social-technical approaches of policy, they might be necessary in order to achieve political aims which have the most priority.

References

Bachmann R., Möll G. (1992), 'Alles neu ...? Rationalisierung von industriellen Innovationsprozessen. Eine Herausforderung für die industriesoziologische Analyse?' in Malsch Th., Mill U. (eds.), *ArBYTE. Modernisierung der Industriesoziologie?* Berlin, pp. 241-270

Berger P.L., Luckmann Th. (1966), *The Social Construction of Reality*, Garden City, New York

Bieber D., Möll G.(1993), *Technikentwicklung und Unternehmensorganisation. Zur Rationalisierung von Innovationsprozessen in der Elektroindustrie*, Frankfurt/New York

Burawoy, M./Krotov, P. (1992), 'The Soviet Transition from Socialism to Capitalism', *American Sociological Review*, Vol. 57, No.1, February 1992, pp. 16-38

Fortescue, S. (1985), 'Project planning in Soviet R&D', *Research Policy*, Vol. 14, pp. 267-282

Fortescue, S. (1990), *Science Policy in the Soviet Union*, London/New York

Fortune, January 25, 1993, *Russia 1993. Europe's Time Bomb*

Freeman Ch. (1988), 'Japan: a new national system of innovation', in Dosi G., Freemann Ch., Nelson R., Silverberg G., Soete L. (eds.), *Technical Change and Economic Theory*, London/New York, pp. 330 - 348

Giddens A. (1976), *New Rules of Sociological Method*, London

Giddens A. (1984), *The Constitution of Society. Outline of the Theory of Structuration*, Cambridge
Glugiewiecz E./Gruchman B. (1988), 'The Role of Innovations in Regional Economic Restructuring in Eastern Europe', in Aydalat P./Keeble D. (eds.), *High Technology Industry and Innovative Environments: The European Experience*, London and New York, pp. 221-232
Kelly T./Salaman M. (1992), 'Swords into Plougshares, Tanks into Telephones', *The OECD Observer*, August/September 1992, pp. 12-18
Lundvall B.-A. (1988), 'Innovation as an interactive process: from user-producer interaction to the national system of innovation', in Dosi G., Freemann Ch., Nelson R., Silverberg G., Soete L. (eds.), *Technical Change and Economic Theory*, London/New York, pp. 349-369
Malsch Th. (1984), 'Erfahrungswissen versus Planungswissen. Facharbeiterkompetenz und informationstechnologische Kontrolle am Beispiel der betrieblichen Instandhaltung', in Jürgens U., Naschold F. (ed.), *Arbeitspolitik. Materialien zum Zusammenhang von Macht, Kontrolle und betrieblicher Organisation der Arbeitt*, Leviathan Sonderheft 5/83. Opladen, pp 231-251
Malsch Th., Bachmann R., Jonas M., Mill U., Ziegler S. (unter Mitarbeit von Hans-Jürgen Weiflbach) (1993), *Expertensysteme in der Abseitsfalle? Fallstudien aus der industriellen Praxis*, Berlin
Marer,P. (1991), 'The Transition to a Market Economy in Central and Eastern Europe', *The OECD Observer*, April/May 1991, pp. 4-10
Mill U. (1992), 'Soziale Anerkennung und Technikgestaltung', in Bergstermann J., Manz Th. (eds.), *Technik gestalten, Risiken beherrschen. Befunde der Sozialforschung zur Entwicklung moderner Produktionstechnik*, Berlin
OECD Economic Outlook (1992), No. 51, June 1992, Paris
Orlikowski W.J. (1992), 'The Duality of Technology: Rethinking the Concept of Technology in Organizations', *Organization Science*, Vol. 3, No. 3, August 1992, pp. 398 -427
Schröder H.-H. (1988), 'Innovationspolitik in der Sowjetunion seit 1985', in Welfens P.J.J./Balcerowicz L. (eds.), *Innovationsdynamik im Systemvergleich. Theorie und Praxis unternehmerischer, gesamtwirtschaftlicher und politischer Neuerung*, Heidelberg, pp. 434-446
Der Spiegel, 12/1990, *Westfirmen heuern Sowjetforscher an*
Smith B.C. (1985), 'Prologue to "Reflection and Semantics in a Procedural Language', in Brachman R.J./Levesque H.J. (eds.), *Readings in Knowledge Representation*, Los Altos
Sviokla J.J. (1986), *PLANPOWER, XCON and MUDMAN: An in-depth Analysis into Three Commercial Expert Systems in Use*, Cambridge (Mass.)
Waterman D.A. (1986), *A Guide to Expert Systems. Reading* (Mass.) et al.

We would like to thank Carmen Pascual Camps for her instructive ideas.

2 Technology organisation and employment – the employment effects of technology and organisational change

Mike Campbell

Introduction

Ever since the 'Luddite' riots in the early 19th century, when weavers, initially in Nottingham, broke up the new wider frames for fear that the increased productivity so occasioned would cost them their jobs in the hosiery industry, there has been popular concern about the effects of technological change on employment. Indeed Ricardo (1821) provided an intellectual echo of popular concern:

'The opinion entertained by the labouring class that the employment of machinery is frequently detrimental to their interests, is founded not on prejudice and error, but is conformable to the correct principles of political economy'.

In essence both popular and academic arguments rest on the simple proposition that technical change is labour saving and hence jobs will be lost as technology advances. In recent years there have been a number of studies which have sought to examine the effects of 'new technology', by which is usually meant microelectronics/information technology, on employment in order to establish the actual/likely extent of job creation or, more usually, job destruction, so entailed.

This paper, after briefly identifying a number of different theoretical approaches to the issue, then outlines a framework with in which it is possible to place empirical studies and which provides a template for them. It then reports on some of the key empirical studies investigating the relation between the application of new technology and employment. The paper then provides an empirical investigation into one key aspect of the relation, the 'direct' and 'output' effects, and a comparison of it to the employment effects of organisational change. The results, inter alia, provide corroboration of the view that even the direct employment effects of the application of new technology have been generally overestimated and that greater attention should be paid to the effects of organisational change on employment. They also provide some useful insights into the debate on the effects of technical change on employment.

Alternative approaches

As is common in the social sciences there are a number of different perspectives, explicit or more usually implicit, in the empirical studies, which inform their investigation. First, the free market, or neo classical, approach argues that the price mechanism will resolve any problems caused by the labour 'displacement' effects of technological change. It accepts, indeed welcomes, the fact, that displacement does occur but the free operation of factor, especially labour, markets and product markets ensures that temporary excess labour supply in re-employed elsewhere in the economy. The 'compensation' effects of expansion in the technology producing sectors, the increased competitiveness of the sector employing the new technology and the increased demand for its products as prices fall under competitive conditions, all occur quickly under market pressures, as does the fall in the price of labour (wage) towards a new equilibrium level as the available supply of labour, temporarily, expands. On this view of the world technology, in the short and long run, has no overall effect on employment. Full employment is ensured by Say's law and the costless, automatic adjustment mechanism of markets. The direct effects are uninteresting in themselves as markets resolve any potential problems.

The market failure approach adopts the above framework but recognises that markets, at both individual and aggregate levels, fail to operate as efficiently as the free market approach suggests. A wide range of lags, inflexibilities and rigidities mean that the compensation effects occur slowly and unevenly - problems of labour mobility, skill mismatch, slow adaptation to 'best practice' technical change and aggregate demand problems, all mean that the direct effects are of interest as market failures ensure that displacement occurs, often substantially, prior to market adjustment and that technical change is capable therefore, for a period of time, under certain conditions, of reducing employment.

The structuralist approach puts the very process of technical change at the centre of analysis. Instead of the focus being on the degree/extent of market failure/adjustment to an exogenous technical change, technical innovation itself drives the system and is inherently unstable. Innovation is seen to come in waves and diffuse unevenly with concentrations in certain sectors. Swarms of innovations accumulate causing the eventual stagnation of innovation. Here the problem for employment is insufficient technical change and a lack of aggregate demand caused partially by innovation stagnation. Again, however, there is then an interest in the direct effects of the innovations.

Marxist approaches focus on the tendency to economic crises occasioned by the rising organic composition of capital (the replacement through labour saving technical change of variable by constant capital) and the consequent tendency of the rate of profit to fall (TRPF). Moreover the process of restructuring itself allows the same number of workers to produce more output, or the same amount to be produced by fewer workers. Hence employment will fall unless output rises fast enough to compensate for the increase output per worker. Again the direct employment effects are of interest from this perspective as there is a 'direct' focus on the restructuring process and the TRPF limits output growth, at least cyclically if not structurally.

A framework for analysis

Each of the above perspectives offers some insights into the relationship between technical change and employment yet none offers a complete account (cf. Alford and Friedland 1985). However these insights taken together provide a framework within which one can locate empirical studies (see Figure 1). Technical change may involve either, or both of, process and product innovations. The former is seen to have a direct effect of job displacement (for a given 'lump' of output) and indirect compensating, effects of job creation in the new technology equipment supply sector and as a result of increased productivity in the organisation(s) adopting the new technology, giving it a cost/price advantage and hence a likely growth in output. The relation between this output effect and the direct effect on employment is addressed in the empirical section of the present paper - does the organisation as a whole shed or gain labour from its process innovations?
Product innovations lend to new products and services being produced for which there is a demand, thus stimulating job creation. Both process and product innovations will diffuse, to some degree and at some rate, through the sector(s) and the economy and will either (net) displace or create jobs depending on the sum total of the direct and indirect effects of process innovations and the job creation effects of the product innovations. The macroeconomic effects of technical change are also important as the aggregate effects on productivity, output, prices, wages, profits, and international competitiveness are felt. As with diffusion the overall effect, a priori, is uncertain as it depends on the degree of market efficiency and the efficacy of government policy as well as overall demand conditions in the economy.

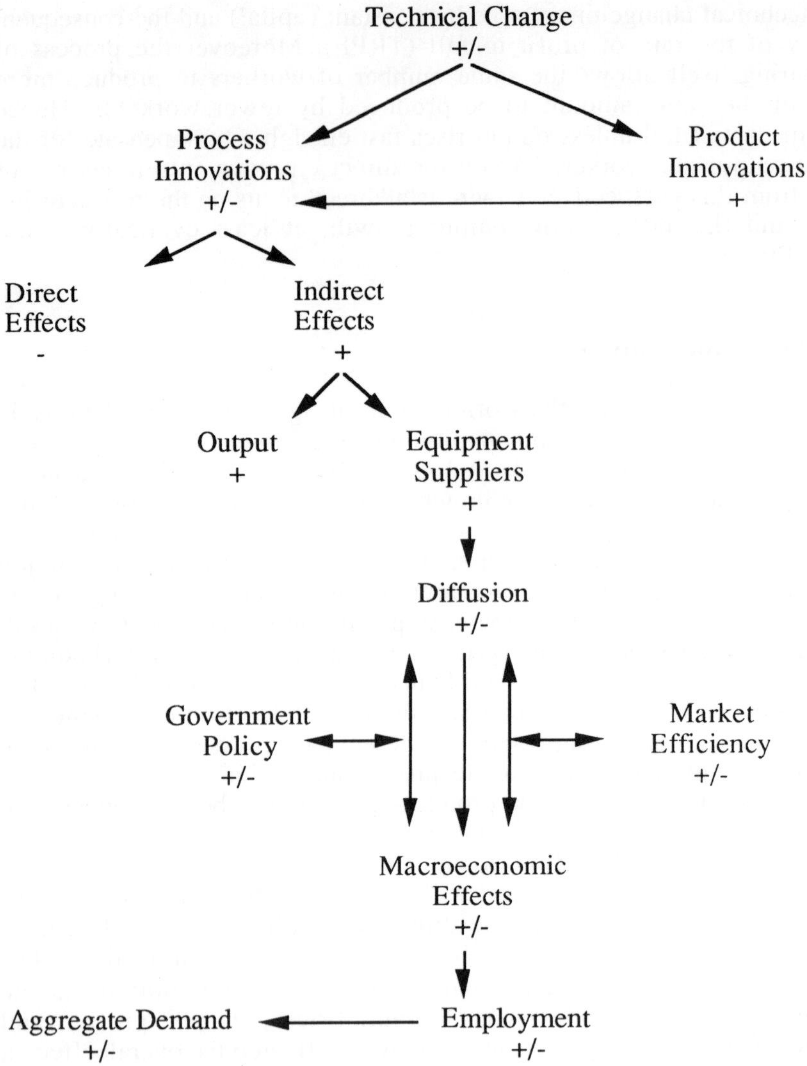

Note: Hypothesised Effects

+	positive effect on employment
-	negative effect on employment
+/-	no a priori view and/or depends on approach

Figure 2.1. The effects of technical change on employment: key components

Some recent empirical studies

The debate about the employment implications of technical change in relation to new or information technology began in the late 1970s. Whilst it is not possible to review all these studies here it is important to draw attention to the range of studies to see how they fit into our framework outlined above. Jenkins and Sherman (1979) estimated that the introduction of IT could reduce employment by 5 million by 2003: equivalent to a 23% reduction in the employed labour force. They also predicted unemployment rising to 15% by 1985. These estimates focus entirely on the widespread and rapid introduction of IT in the process of production in manufacturing and services and wholly on its direct effects. Rothwell and Zegveld (1979) argue that the 'jobless growth' in manufacturing in the 1970s was primarily due to the employment effects of the introduction of labour saving technologies increasing capital intensity. Northcott et al (1985) on the other hand estimated that IT related jobs loss had only been 0.5% of total employment in manufacturing and less than 5% of the loss of jobs from other causes. This was based on a survey of 1200 factories and what had actually been the experience of introducing IT. In two thirds of all cases of the introduction of microelectronics there has been no change in the level of employment in the organisation as a result. The focus here, then, is on the direct and output effects at the micro level - similar in that sense to our empirical approach below. Taylor et al (1985) in a local study of a London borough in somewhat similar vein found that whilst employers 'expected' 1 in 5 jobs to be affected by the introduction of New Technology, only between 10-20% of these would be a job loss effect i.e. between 2-4% of total employment. Batstone and Gourlay (1986) in another survey found that half of respondents reported a 'negligible' effect of IT on employment, whilst 1 in 4 reported that it had been associated with a loss of jobs of 10% or more. Whitley and Wilson (1986) adopted a quite different approach - modelling the macroeconomic effects of the introduction of IT and in principle, as far as highly aggregate economic models allow, taking account of the key components in our analytical framework above. They forecast a 'modest' decline in employment between 1985-1995 due to IT of 1%. The small size of the net loss was due to the compensating effects of the increased competitiveness and improved international trade performance that resulted. Wobbe (1989), in a much more specific and detailed study of the effects of robotics on employment, estimated the direct effects at a loss of 120,000 between 1980-95, but total compensation effects of all the other effects at 90,000, giving a net job loss of 30,000 or 0.8 of a person for each robot introduced.

In a study very much in the structuralist mould Freeman and Soete (1987) develop a sophisticated analysis based on their TEMPO research programme which cannot be done justice to here. Suffice it to say that they conclude that 'new technologies could play a very important part in generating new employment opportunities' and that the combined effects of the compensation mechanisms (productivity, prices, output, income, new products and so on) swamp the direct effects of labour saving process innovations. Furthermore they show that the employment problem arises not from too rapid a growth of productivity, occasioned by the application of new technology to process innovations, but from too slow a growth of productivity occasioned by a lack of process innovation. Nevertheless this does not imply that they adopt a free market approach, for the latter has occurred precisely because of market failure, the uneven diffusion of innovations and limited product innovation, low rates of growth of aggregate demand, inappropriate government policies and a tendency to undertake process innovations for rationalising (restructuring to reduce unit labour costs and capital deepening) rather than expansionary (capital/widening and increases in output) purposes. They also found that the only manufacturing industries to increase employment were precisely the microelectronics based sectors.

However they did find a significant 'break' in the relation between output and productivity growth (the so called Verdoorn relation) in the mid 1960s; so that whereas previously a 2% rate of growth in output was required to maintain employment, in the late 70s and early 80s this had grown to 3.5%, thus implying a shift in the balance of effects of process innovation from output to direct effects.

Perhaps the most extensive, comprehensive and recent study is reported in Christie et al (1990). 1200 manufacturing establishments were surveyed in 1981, 1984 and 1987 with a view to establishing, inter alia, the employment effects of the introduction of microelectronics in both production processes and products. By 1987 75% of factories were using microelectronics (double that in 1981). We focus here on the direct process innovation effects. Overall the ratio of job losses to gains had increased over the period from 28,000 in 1981-3 to 82,000 in 1985-87. However to put this in context, it is equivalent to just over 2 jobs per factory per year and 52% of factories reported no change in numbers employed due to microelectronic applications at all. Nonetheless multivariate analysis shows that the greater the use of microelectronics the greater the job loss especially in food, drink and tobacco, chemicals, vehicles, printing and mechanical engineering sectors as well as in large firms overall. In small firms (less than 20 employees) the effect on employment was positive! Overall the number of job losses due to the direct effects of process innovative was very small (gains and losses) compared to the overall job losses in the factories through the 1980s. They conclude that while new technology is not the 'job killer' that some suggest, it does and has, allowed large productivity and output gains to be achieved without significant employment growth, a finding from micro analysis that parallels the aggregate study of Freeman and Soete (1987) and implies that there will be little if any growth in the level of employment in manufacturing in the future despite increases in output.

An empirical study of new technology and organisational change in Yorkshire and Humberside

Introduction

Our study displays a number of distinctive features which merit its consideration. First, it is based on face to face interviews with employers. Second, the questions on new technology are embedded in a wider study thereby not overfocusing (estimating?) on them. Third, the sample survey is representative of the pattern of industry and employment in the region. Fourth, it provides a detailed regional focus. Fifth, it covers both service industries and manufacturing. Sixth, it is more recent than the studies reported here, being undertaken in 1991. Finally, and most importantly, it compares the employment effects of technical change to the employment effects of organisational change. The focus is on the direct and output effects of process innovations (not just on the 'shopfloor' but throughout an organisation) and thus investigates the overall employment effect at a place of work and the balance between the two dominant employment effects of process innovations.

Methodology

A series of face to face interviews were undertaken with a representative sample (by location, size and industrial sector) of companies in Yorkshire and Humberside in 1991 based on a pre-prepared interview schedule (see Appendix). The interview survey was designed to obtain a wide range of employer based information including establishment and workforce details, employment change, occupational structure, recruitment difficulties, skill shortages, training and, of central relevance to this paper, use of new technology and organisational change. The questions used appear as sections F and G in the interview schedule in the appendix.
The survey and coding was undertaken by the region's Training and Enterprise Councils under contract to the Employment Department and often subcontracted to market research or academic research units. The Policy Research Unit itself undertook several hundred of the interviews. The survey data was input by the Employment Department and was obtained by the Policy Research Unit from them primarily in order to undertake a detailed study of recruitment difficulties and skill shortages (Campbell and Baldwin 1992). Data from each section of the interview schedule was held in separate files and knitted together in SPSS-X using a merge and sort process. Once the data manipulation was complete it was analysed using SPSS-X.

Sample characteristics

The sample consisted of 1951 employers, employing nearly 271,000 people in all, or 14.2% of employees in employment in the Yorkshire and

Humberside region. It was distributed in the following manner across the regions 9 Training and Enterprise Council (TEC) areas (see Table 1).

Table 2.1
Employers by TEC area

TEC	No. of Employees
Barnsley/Doncaster	400
Bradford	174
Calderdale/Kirklees	121
Humberside	145
Leeds	271
Rotherham	133
Sheffield	250
Wakefield	213
North Yorkshire	244
Yorkshire and Humberside	1951

Table 2 shows the distribution of the sample across size bands of employers, so that, for example, 45% of firms employed less than 25 employees. Table 3 column 1 shows the distribution across sectors by standard industrial classification so that, for example, just less than 13% of firms were in the engineering sector. Column 2 shows the actual distribution of employers across the region. We can see therefore that the sample is broadly representative of the regions' industrial structure.

Table 2.2
Employers by size of firm

No. of Employees	% of Total
1-24 Employees	45
25-199 Employees	39
200+ Employees	16

57% of employees in the sample firms were male and 43% female, whilst 79% of the workforce worked full time, 21% worked part time.

Table 2.3
Employers by industry type

	SIC Sample	% of Total Population
Agriculture, Forestry and Fishing	1.5	1.4
Energy, Water	2.1	2.9
Metals/Chemicals	3.6	4.7
Engineering	12.9	9.4
Other Manufacturing	11.9	12.3
Construction	5.7	5.3
Distribution, Hotels and Catering	27.2	21.4
Transport and Communications	5.5	5.4
Banking, Insurance and Finance	21.0	29.0

The occupational structure of employment in the firms surveyed is given in Table 4, so that, for example just less than 21% of employment was in the clerical/secretarial group.

Table 2.4
Occupational structure of employment

SOC	% of Total
Managers	20.3
Professional	7.5
Skilled Technical	6.1
Clerical/Secretary	20.7
Skilled Manufacturing/Craft	11.1
Personal/Protective	6.2
Sales	8.6
Plant Operatives/Semi Skilled	9.0
Unskilled	10.3

Key results

Seventy eight percent of the firms were currently using new technology, across a wide range of areas of work (see Table 2.5).

Table 2.5
Current use of new technology by area of work (%)

Design	12
Materials/Manufacture	15
Storage, Packaging and Distribution	15
Word Processing	38
File Storage/Processing	23
Communications	31
Finance/Accounts	38
Salaries/Personnel	29
Data Analysis	26
Other	9

However in 90% of the instances of the introduction of new technology there was neither an increase nor decrease in the number of jobs in the firm. Indeed out of the 1600 or so firms who had recently introduced new technology only 48 reported resultant job gains and 36 reported job reductions. In only 8 cases were more than 5 jobs lost. In all, 172 jobs were gained and 266 lost - a net loss of 94 jobs. To set this in context the 1951 organisations surveyed employ 271,000 people.

In terms of which occupations gained and lost jobs, the key gainers were clerical/secretarial; professional and skilled technical groups. The key losers were clerical/secretarial; sales and semi skilled manual. It is interesting that the same occupational group can, depending on the organisation, be a major looser or beneficiary of the application of new technology, implying that factors other than the technology per se are central to understanding 'its' employment effects.

In terms of industry the main gainers in employment terms were engineering, metals/chemicals, construction and 'other services' whilst the main losers were metals/chemicals and transport/communication. In general it was the larger firms who gained most jobs and lost most jobs, though medium sized firms also experienced job increases.

It is important to be clear: these occupational, industrial and size patterns of job gains/losses refer to a very small proportion of cases of the introduction of new technology. In most cases there is no discernible effect in either direction. However where changes do occur, in general, both gains and losses are experienced, implying that the organisational, market and other contexts in which it is introduced is as important in determining its employment effect as the new technology per se.

When firms were asked identical questions about the employment effects of organisational change (managerial changes, changes in structure, shift patterns, ownership and so on) just less than one third (30%) had recently experienced such changes (compared to nearly four fifths who were currently using new technology). In over 70% of cases there was neither a resulting increase or decrease in jobs in the firm as a result. This compares to 90% of cases in the case of changes in technology. Of the nearly 600 firms who had undergone organisational changes 120 experienced job gains and 62 experienced job losses. In all 507 jobs were gained and 741 jobs lost - a net loss of 234 jobs. So, whilst the impact of organisational change on total employment is, as with new technology, relatively small, it is nevertheless significantly greater, especially at the level of the firm itself, as many fewer firms experienced the former yet the employment effect is much greater. Table 6 summarises the position.

Table 2.6
Employment effects of new technology and organisational change

	New Technology	Organisational Change
% of firms experiencing	78	30
% of those firms with no employment effect	90	70
Nos. of jobs gained	172	507
Nos. of jobs lost	266	741
Net job loss	-94	-234

The average gross job loss per firm affected by new technology is 0.16 of a person whilst the average gross job loss per firm affected by organisational change is 1.23 persons. Again, however, in aggregate the effects of both technical and organisational change are very small when compared to the pattern of employment gains/losses in the region as a whole during the year. The effects of the level of demand for goods/services and competitive pressures swamp the effects of both technical and organisational change both at the industry and aggregate regional level. The finding is consistent with Freeman and Soete (1987) and Christie et al (1990).

In terms of the occupational employment effects of organisational change Table 7 compares these changes with that induced by technological change.

Table 2.7
Occupational employment effects of new technology and organisational change

SOC	% Distribution of Job Gains/Losses by SOC[a]	
	Gains	Losses
Managers	8 (0)	3 (0)
Professional	13 (10)	0 (13)
Skilled Technical	13 (9)	0 (9)
Clerical/Secretarial	42 (6)	69 (3)
Skilled Manufacturing	7 (15)	0 (21)
Personal	2 (18)	0 (15)
Sales	2 (18)	11 (2)
Plant Operatives	13 (16)	11 (13)
Unskilled	0 (15)	6 (18)

[a] The % in brackets ( ) refer to Organisational Change, those without brackets, Technological Change. The figures refer to the % of the number of instances of gains/losses, rather than the absolute numbers of jobs gained/lost. So, for example,. it means that 8% of all instances of job gains through technological change was amongst managers.

A number of important conclusions can be drawn from these results. First, the job losses associated with organisational change have a more widespread effect across occupations than technological changes. Second, job losses associated with technological change very much focus on office rather than production jobs, with no discernible effect at all on skilled technical or manual workers, professional staff or personal service staff. Job losses associated with organisational change, whilst widespread, do affect most skilled manual workers, unskilled manual workers, personal service workers, plant operators/ semi skilled workers and professional staff.

In relation to job gains, again there are large differences between the occupations affected most by technological and organisational change, with in the former case gains focusing, broadly, on 'white collar' jobs, and in the latter case gains focusing on 'blue collar' sales and personal services.

In terms of the industrial distribution of employment change associated with technical and organisational change, the position is summarised in Table 8.

Table 2.8
Industrial employment effects of new technology and organisational change

	% Distribution of Job Gains/Losses by SIC[a]	
	Gains	Losses
Agriculture, Forestry and Fishing	30 (75)	20 (16)
Energy/Water	0 (32)	0 (33)
Metals/Chemicals	23 (18)	30 (0)
Engineering	23 (18)	8 (37)
Other Manufacturing	8 (25)	6 (0)
Construction	22 (70)	0 (6)
Distribution, Hotel and Catering	4 (36)	8 (22)
Transport/Communication	10 (44)	25 (11)
Banking, Insurance, Finance	7 (28)	7 (33)
Other Services	13 (58)	5 (14)

[a] The % in brackets ( ) refer to Organisational Change, those without brackets, Technological Change. The figures refer to the number of instances of gains/losses rather than absolute numbers. The figures in this table should be interpreted as follows: 30% of all organisations who had jobs affected by new technology in agriculture experienced job gains.

Again there are a number of important observations one can make on these results. First, the pattern of job losses across industries associated with organisational change differs markedly from that associated with technical change, with particularly marked differences in energy/water; engineering; distribution; and banking, where losses due to the former predominate. Losses due to the latter predominate in metals. chemicals and transport and communications. Second, the pattern of job gains across industries associated with organisational change differs markedly from that associated with technical change. The most marked differences where gains due to the former predominate are in agriculture; energy; other manufacturing; construction; distribution; transport; banking and other services. Gains due to technical change predominate in metals/chemicals and, to a lesser extent, in engineering. Third, in most industries the organisational affect outweighs the technical change effect with regard to both gains/losses.

Table 2.9
Employment effects of technical and organisational change by firm size

	% Distribution of Jobs Gained/Lost by Firm Size[a]	
	Gains	Losses
Large	13 (16)	14 (29)
Medium	10 (46)	8 (20)
Small	10 (49)	3 (13)

[a] The % in brackets ( ) refers to organisational change, the % without brackets, technical change. The figures refer the % of firms, of a particular size, who had been affected by technical/organisational change, who had experienced job gains/losses. For example 13% of large firms who were affected by technological change experienced job gains.

Finally we can compare the employment effects across different sizes of firm: small (0-19 employees), medium (20-199 employees) and large (200+ employees). The position is summarised in Table 9. We can see that with regard to job losses, and gains, organisational change has a greater effect in all sizes of firm though the dominance is greatest in medium and small firms and with regard to job gains. The employment effects of technical change, though less than that of organisational change, is relatively great on large firms, particularly in relation to job gains.

Concluding remarks

This paper has identified a number of different approaches to the examination of the effects of new technology on levels of employment; offered a framework within which empirical studies of these effects can be placed, so they can be more coherently positioned in the literature, and which provides a conceptual insight into the issues at stake; reviewed recent empirical work in the light of this framework; and reported on a recent large scale empirical study based on primary data, on the direct/output effects of technical change on employment within organisations and compared these employment effects to those occasioned by organisational change.

It is hoped that the paper has offered a number of insights into the debate on the relation between employment, technical change and organisational change and that the empirical results make a further contribution to the debate. In particular it has demonstrated the relatively minor employment effects of both technical and organisational change and the relative dominance of the latter, together with further results showing the pattern of these changes across different occupations, industries and sizes of firm.

References

Alford R and Friedland R (1985), *Powers of Theory*, CUP

Batstone E and Gourley S (1986), *Unions, Unemployment and Innovation*, Blackwell

Campbell M and Baldwin S (1992), *Recruitment Difficulties and Skill Shortages*, report: for the Employment Department, Policy Research Unit

Christie I et al (1990), *Employment Effects of New Technology in Manufacturing*, Policy Studies Institute

Freeman C and Soete L (1987), *Technical Change and Full Employment*, Blackwell

Jenkins C and Sherman B (1979), *The Collapse of Work.*, Methuen

Northcott J et al (1985), *Chips and Jobs*, Policy Studies Institute

Ricardo D (1821), *Principles of Political Economy* 1951 edition, Cambridge University Press Rothwell R and Zegveld W (1979) Technical Change and Employment. Frances Pinter

Taylor A et al (1985), *The Impact of New Technology on Local Employment*, Gower

Wobbe W (1989), *The Employment Effects of Robots and Flexible Manufacturing,* FAST Occasional paper No 218, EC, DG 12

3 Organisational implications of technological change. Aspects of dynamic forms in applied systems

Savvas Katsikides and Karin Schneider

Abstract

The major changes currently taking place in organisations and technology are creating new research areas and questions for the traditional thinking in social sciences. Of emerging importance for macrosystems, are culture, production, organization, human potential, and of course technology. This paper is an examination of the structures within organisations, which lead on from Hardware to Humanware. It is an introduction and deciphers a discourse from the manufacturing to the post-industrial system. From the hard and software production systems to the anthropocentric approach (human centred systems).

Introduction

The development and extension of enterprises, both on a national and international level, brought about many changes in the hitherto organizational structure. The core of this paper is to attempt to throw light on the relationships between industrial manufacturing and technology and on the other side the emerging anthropocentric approach. The continually changing organisational structures, for example due to the new technological approach or relocation of the production plant, prove that a paradigm change of the enterprises' conception is underway. The evaluation and comparison of these above mentioned systems will be presented in this article.

The organizational implications were always a characteristic of the adaptation of technology. Advances in technology and organization recreate the manufacturing and management system and exclude the non compatible structures by means of human manpower and organization. Questioning concepts dealing with how organizations are now functioning becomes not only a scholastic term for effective management, but for the first time shows serious implications in the practical daily use of computer supported technology. Implementation of technology, requires more skilled workers and judgement, initiates new job qualifications which are based on technological innovations, eliminating, on the other hand, the non up-dated compatible human capital. The aim of this paper is to analyse the inter-organizational relations and aspects of technology adaptation processes with respect to innovations which bring about drastic changes in the routine working procedures.

Definitions and framework

Rapid technological changes and their applications in many sectors of the economy demand and create qualifications which can be extended over the life cycles of hardware and software. To a much greater extent, qualifications which emerge among the administration of processes are system eminent. which implications does this have in high-tech industry and in the service sector like hospitals? How can new knowledge both be adopted and be skilled? What is the current state of these qualifications which are shorter than the life cycles of hard and software? What is the common point between profit oriented organizations and those which are offering social services as part of the welfare state, such as hospitals?

The classical argument that the organisational structures of enterprises could be seen as constant can no longer be seen as acceptable, since other parts of the respective life of an enterprise incorporate more and more external parameters. Whilst talking about how technical pictures emerged, Flusser argued in 1985 that they can be transferred in a symbolic way, and in our case, into the classical acceptance how the enterprises' environment has developed. Connecting this with the working environment we can detect three main phases, namely: a) the embedding of huge masses of humans into the working process, in which the work can be seen as a constant and the working environment as a variable. The variables were the tools and the executive human power. b) the second phase was developed and shaped based on the dynamics of the first process and converted the human to a variable. The human was the executive power, of course dependent on the constants of the enterprises´ organisation. The systematic Taylorism, as well as other management and organisation theories, had standardised the inner life of the enterprise. c) the third phase which is supported by a maximum technological equipment, takes with itself humans and tools in the position of variables. It left the continuously changing organisation, the macrosystem, to look like a constant. According to that, the organisational structures of an enterprise follow a dynamic model, in which permanent compatibility allows the increasing mechanisation to set root in the firm. As an example, the highly-computerised office forced the various activities towards a certain standardisation. Wobbe (1986) emphasized this by saying that:

"EDP is a way of organisation or of organising technology. At the same time a political discussion emerged due to the creation of employment generated by that, for during the process of mechanisation and in particular by organisational changes, different groups of workers and working interests will be automatically affected."

The social power of the macrosystems is very important since new organisational structures still incorporate the whole view of the market, technology, and personnel, so that strategic competition can build up further. The transformation of the enterprise strategy is successful within the human component. Consequently, the human function is responsible for success and establishes one of the important sources of enterprise strategy. An organisation is compatible by these means or successful if it is not static, or, if we go back to Tuerk(1989, p. 25) organisations are only rudimentary, and stable only in the long term, it is more usual to perceive them as being permanently in motion; they never reach their surface stability, neither by balance nor static, but only by movement. New organisational images produce a constantly changing social reality because their duties, due to the changing market variables, must be redefined every time. On the other hand, these processes head for the working class with determined effect.

Much attention has been paid in the past years to the so called "effective implementation of technology" (Adler and D. Helleloid, 1986, Majchrzak, 1988), or the meaning of local knowledge systems in advanced technology organizations (Baba 1990). On the other hand, approaches like skill-based design which leads to organizational effectiveness are becoming more and more respected when considering how to turn technologies into tools (Salzman 1989), or, organizations using symbolism as an emerging asset of the creation of realities within the firm, based upon their own professional histories, personalities, value systems[1] (Astley, 1985) etc. Furthermore, the aspect of the social construction of technological systems (Pijker et al 1987) arose among technologists, and many more people now know that culture, language and ideology are definitely terms of management. Uncertainty of technological systems has the same effect on organizational concepts, because organizational structures should become more organic and less mechanistic (Kolodny 1990). Flexibility, on the other hand, replaces efficiency as one of the key objectives of both the manufacturing and service sectors.[2]

As the courier of the new technological system, the hardware sector, after viewing the evolution in the software sector and loosing its monopoly, decided to catch up using flexible and user friendly systems in order to be more accessible to the user. Small and medium sized industries were more flexible and quicker to react than the larger manufacturers (Silicon Valley Fever).

Introduction to the developments in computerization in the health sector: the hospitals' approach

The health care sector has increasingly received political attention at the international level because the financing of currently valid health insurance systems has reached its limits. The first consequence of this evolution was to find a more "managerial approach" in leading health care institutions with the aim of reducing costs through higher efficiency and rationalization policies. With the maturing of information technology and its successful introduction into the office environment, the health care organizations have turned their attention to the innovative potential of computers as a competitive weapon. On the one hand, information technology perfectly meets the enormous information needs in patient care. On the other hand it is viewed by hospital managers as the ideal support for documenting and thus controlling the production process in hospital industry which is exemplified in the introduction of patient classification systems (Pouvourville, 1985).

The objectives for information systems technology are, however, more complex than the simple generation of activity reports. An integrated information system is expected:

- to provide a balance between antagonist groups within the organisation. Especially the medical and administrative powers strive to intervene actively in the development process to preserve their interests,

- to co-ordinate the relations between the medical-technical platforms and the medical services, which are characterized by heavy workloads induced by slow means of communication,

- to simplify the work organization within the wide variety of different functional areas,

- to improve the patient accounting procedures,

- to provide the ultimate integrated documentation support in the form of the computerized medical record and many more desirable properties of equal importance.

Many hospital institutions have adapted information systems for the needs of the administrative department concerns. The type of system which best fits their requirements were mostly centralized architectures with solutions that have been imported from other large organizations forms, such as those which can be found in insurance companies. Until recently, co-ordination problems between the multiple participants involved in patient care as well as the diversity of their specific information needs have been the principle obstacles to the computerization of the medical field.

Current system developments are mainly focused on information retrieval problems related to patient care. The patient record traditionally is used as primary working instrument. It can be defined as a unique, individualized reference document. In any case, it plays a central role in the negotiation of computer development between hospital administrators, computer engineers and the actors participating in patient care. The computer system reflects to a large extent the need for **documentation** and **communication** through patient record manipulation tools.

The organizational environment of computer applications favours integrated computer architectures which emphasize the administrative control of patient movements within the hospital. The operational systems are, to a greater extent, equipped with modules for:

- *the automization of laboratories ,*
- *support for personnel planning ,*
- *the maintenance of administrative patient data bases ,*
- *scheduling of consultations and examinations,*
- *accounting tasks ,*
- *automatic registration/control of medical orders,*
- *patient management.*

Whilst the realization of permanent computer archives is technically feasible (with the memory capacity of current technologies having become economically attractive), the choice of system development methodology as a management tool has become a critical success criterion. The mutual dependency between system development teams and user groups for the construction of well designed systems has started an intense discussion on the definition work content, and appropriate evaluation methods that will emphasize confidence in the system. However, in many cases, the disorientation on the one and only right way to develop, and the lack of experience in this new market, contribute to the overall impression of disorganization with respect to the introduction policies.

The case studies

The hospitals analyzed in the survey were selected from the public and private sector. For the most part, they were attached to a university teaching hospital centre. All institutions had made, or were planning to make, investments in computing equipment which besides handling financial management tasks, could also provide support for the organization of a patients' hospitalization from the perspective of the medical-technical professional or the clinician.

The study focused on the development of the information system, and acquisition policies applied affecting the main characteristics of the target system, as well as the user involvement. Four hospital institutions are discussed. The first case is a paediatric city clinic which runs with an integrated hospital information system that was developed externally by a major computer manufacturer. The second case is a private institution for the treatment of cancers and cancer related diseases, equipped with an internally developed system. The third case is the surgical department of a cardiological hospital operating with a highly complex local system combining the time-critical demands of the ICU setting with the general care requirements of pre- and postoperative treatment in the ward. The fourth case is a regional hospital centre maintaining a loosely coupled decentralized architecture of local networks.

The pediatric city hospital

The publicly managed general city hospital was recently opened, and joined two older hospital institutions. A separate data processing department handles all information system related tasks. All wards within the hospital are connected to a central main frame unit. The laboratories are controlled by a separate system linked with the main unit, thus providing total transparency at the user level. The medical secretarial offices are in addition equipped with personal computers. The major requirement for this integrated computer architecture was the optimization of communication between the functional units. This mainly concerns the transfer of documents, laboratory samples, test results, or medical orders. Especially the wards, the laboratories and the administrative units have been included in the concept. A single software package controls the relations between units and the organization of work within a unit. A patient management module co-ordinates patient hospitalization, from the admittance in the patient admittance office to the discharge. The medical orders system dispatches requests and delivers the results accordingly.

User participation in the design and selection procedures of the equipment was practically inexistant, with the exception of the chief executives from the medical domain, and the hospital direction which were present in the mixed working groups, gathering experts from the supplier company and the government. The original proposed solution had to be adjusted to the local requirements of the French health system. Even with a product that had been approved on the American market, major amendments had to be carried out. The strong "costing out" design orientation was considered by the users as a primary obstacle, since it is not part of the French working culture in the health care environment. Besides this fundamental issue, the medical professionals had to familiarize themselves more than any other professional group with the system. The orders module forces them to fully exert their prescription role exclusively using the computer tool. Recognition of this change in the working procedures demanded a considerable time and adaptation effort.

The essential difficulty was the poor flexibility of the data processing department in complying with user requests. As a consequence, all user groups organized themselves into working groups. Meetings with the data processing staff are scheduled on a regular basis to discuss proposals or bring forth problems, experiences in the daily work with the computing system. This feedback brought about a significant change in the quality of the application software in use. Although the conceptual foundation cannot be questioned anymore, the concrete interventions of the hospital staff has changed the practice in the selection, modification, design and implementation tasks in the sense, that the affected users intervene already in the design stage. The introduction of this new technology has also brought up resistance among the other professional groups such as the nurses who directly depend on the physician's prescription role.

The city cancer clinic

The privately managed cancer centre in one of the major French cities is composed of 26 different medical units and technical departments, most of which are connected to a central computing system. In comparison with the paediatric city hospital, the size of the computer expert team in relatively modest (one computer system engineer assisted by two full-time and three part time system analysts). The introduction of the computing system was motivated by a general discontent with the inefficient practice in the maintenance of paper record.

In this institution the system design and implementation process was controlled by the medical staff. The project leader was the medical director of the institution. A co-operation contract with a major supplier of hard and software equipment was made. The computer manufacturer provided a requirements specification and a basic software environment for the generation of a data base. A low cost solution with minimal architectural disruptions was set down.

The first application for the management of medical records was then written by the local computer professional (who was hired at that time). The presently used system consists of a main frame unit installed in wards, and offices. The major demand on the part of the clinicians for fast access to patient related information was the main functional requirement. The central application module is therefore the medical record which is structured into a set of folders, typified by the professional group, the medical discipline, and functional areas within the hospital. The design of the computerized record was modelled on the paper record.

Access to medical records is connected to organizational tasks, such as patient admission, discharge, transfer, as well as tasks related to the generation of medical reports or orders. The user interacts through a set of functions and sub folders defined for his profession. The system is well adapted to the medical and secretarial work environment. Gradually modules have been added for the laboratory assistant, and the administration clerk.

Although the implementation work has partly been delegated to the supplier company, the design and implementation policy was one of formal control exerted by the representatives of the hospital institution. Agents from the technical, the laboratory, the secretarial, and the medical staff were always present in the mixed commissions, negotiating with the supplier. They also met (and still do at need) in the event of major changes to the existing hard and software environment. Reactions of refusal with respect to the technology in user were relatively infrequent, due to the general consensus on sharing information with electronic devices and the precautions taken to develop application which are exactly tailored to the needs of this specific institution. At times, user participation also takes place informally in daily problem situations where the data processing professionals are contacted and informed about errors, requests, improvements etc.

The cardiological clinic

The cardiological clinic maintains a surgical department with two units managing pre and post operative care in the wards, an OP, two intensive care units and a local laboratory. The surgical units are controlled by a local computing centre. Patient treatment spans from the preparation to an intervention, through to a 48 hour supervision in the ICU and the post operative case in the ward.

A single computer expert is responsible for all system acquisitions, maintenance, design and implementation tasks. The present computing architecture evolved from an isolated monitoring system to a highly complex architecture integrating the collection of real-time data (vital signs of the patient) delivered during the postoperative period in the ICU with the medical and especially the nursing practice in the wards. Every stage of the hospitalization is documented in detail on a single patient record.

On the hospital institution scale the system in this department is a decentralized solution, whereas on the departmental level, it is a centralized system. A hybrid central unit (designed for batch, and real-time multi-user work environments) controls the communication between the wards, the ICU, the OP, and the laboratory. The system was originally purchased to handle real-time processing of data in the ICU. Its logic is therefore centred around the efficient storage and display of short-lived data.

The functionality was later extended to integrate the physician's prescription role, the nursing care process, the actions of the physiotherapist, and the pharmacy stocks. The nursing documentation system is a particularly developed feature. It is entirely based on the knowledge and experience of the nursing staff in the surgical department. The experienced representatives from the medical and paramedical staff participated in sketching out the basic design. In the realization of the computer application, the computer expert was, in turn, allowed complete autonomy.

The outcomes of his implementation work were presented as prototypes to the users. They were repetitively modified until final acceptance by the staff.

From the early design stages on, the affected users were intricately involved in the design decisions. Their contributions with respect to the semantics and the presentation of the record, were used as the primary source for the requirements specification. The system is widely accepted and heavily used. Its success is based on the direct and informal relationship between the computer expert and the hospital staff on one side, and on the integration of multiple perspectives into the well comprehended computerized record.

The regional hospital centre

The regional hospital complex is structured into a set of distinct buildings arranged in a campus area. In the computerization process these architectural conditions were prioritized. The computing system in use is a decentralized cluster of local networks connected to a central interface component which handles the exchanges between a main frame unit used by the administration and medical units. Presently two thirds of the departments are computerized.

Each unit is equipped with a local network of personal computers with one master station and a printing device. The overall software platform is based on two separate applications. The first is an integrated package treating matters of the hospital administration. The second application is centred around the patient record and is embedded in standardized hard- and software environment. Patient records are stored on a commercially available data base system. Information is assigned to three categories ranging from a collection of publicly available hospitalization summaries, to specialized records owned by units and individual physicians.

From the initial design stages on, the future users have been involved in the development process, which was managed by the local computing department within the hospital institution. They intensively contributed in creating the specialization records concept. Their constant participation is an absolute necessity to maintain a regular system operation.

Especially with respect to the local data organization and archives the units are totally independent from the data processing staff. The local data processing experts only intervene in problem situations. The resulting decentralized organization therefore also brings about an equal distribution of knowledge going beyond the simple manipulation of a computer programme. In each unit a local system administrator from the medical or paramedical profession has been specially trained to maintain the network. The users can thus seek help from this internal communication partners.

With a special record structure tailored to the requirements of each medical discipline, the basic information needs of the physician has been satisfied. However, a major source of discontent and poor overall work performance with respect to the nursing profession, the improvement of the relations with the laboratory sector has gone unnoticed. With the current system functionality only two professional groups (doctors and medical secretaries) have been integrated. The interests of the vast majority of the hospital staff have unfortunately been ignored.

Conclusions

In general, this investigation of computerization processes in French hospital institutions has shown that centralized and distributed computer architectures are in daily use. The advantages of ready made, externally developed and strongly integrated computer architectures are still more appealing to hospital managers, than loosely coupled system configurations whose complexity is technically and socially more difficult to control.

The administrative and medical professional groups have by far been more involved in the design and implementation of computer systems. The question how paramedical, especially nursing professionals are to be integrated in an overall computerization process remains largely unanswered, one of the main reasons being, that physicians still are doubtlful about a systematization of nursing work knowledge.

On the historical level, the introduction of new technologies in the clinical working environment is so unique in each of the hospital institutions visited, that direct comparisons are difficult to make. Experiences gained with computing systems in one hospital institution can therefore not be generalized to serve as a normative reference for the system design, in other hospitals. Therefore tight costumization seems to be the most powerful determinant of successful computer applications. Yet, in an organizational structure that so strongly reflects the professional hierarchies, especially in the working relationships between nurses and physicians, total integration on a routine basis is a long term objective.

Notes

1. Zahrly, Jan (1990), 'Organization Design', in L.R. Gomez-Mejia/M. W. Lawless, *Organizational Issues in High Technology Management*. Jai Press Inc., pp 79
2. Kolodny, H.F.(1990), 'Some Characteristics of Organizational Designs in New/High Technology Firms', L.R. Gomez-Mejia/M. W. Lawless, *Organizational Issues in High Technology Management*. Jai Press Inc. pp 166

Refereces

Adler, P and Helleloid, D. (1987), 'Effective Implementation of Integrated CAD/CAM: A model', *IEEE Trans. Eng. manag.*, vol. EM-34, no. 2.

Astley, W.C. (1985), 'Administrative Science as socially constructed truth', *Administrative Science Quarterly*, 30, 497-513

Baba, L. Marietta (1990), 'Local Knowledge Systems in Advanced Technology Organizations. in Organizational Issues' in *High Technology Management*, L. R. Gomez-Mejia, Michael W. Lawless, Jai Press Inc. Greenwich, Connecticut, PP 57-75

Bijker, W., Hughes,T.P., Pinch, T. (1987), *The Social Construction of Technological Systems. New directions in the Sociology and History of Technology*, Cambridge/Mass.-London.

Curtis, Bill, Krasner Herb, Iscoe Neil (1988), 'A field study of the software design process for large systems'. *Communications of the ACM*, Vo. 31, Nr. 11. pp 1268-1287.

Downey, Kirk H., Brief, Arthur P. (1986), 'How Cognitive Structures Affect Organizational Design: Implicit Theories of Organizing', in Henry P. Sims, Jr., Dennis A. Gioia and Associates, *The Thinking Organization: Dynamics of Organizational Cognition*, pp. 165-190. Jossey-Bass Publisher.

Flusser, W. (1985), *Ins Universum der technischen Bilder*, European Photography.

Gioia, Dennis A. (1986), 'Symbols, Scripts, and Sensemaking: Greating Meaning in the Organizational Experience', in Henry P. Sims, Jr., Dennis A. Gioia and Associates, *The Thinking Organization: Dynamics of Organizational Cognition,* pp. 49-74. Jossey-Bass Publishers.

Lincoln, Yvonna S. (1985), *Organizational Theory and Inquiry: The Paradigm Revolution,* Sage Publications.

Majchrzak, A et al. (1986), 'A quantitative assessment of changes in work activities resulting from computer assisted design', in *Behaviour and Information Technology*, vol.5. no. 3

Pouvourville, Gerard (1985), 'Hospital System Management in France and Canada: National Pluralism and Provincial Centralism', in *Sc.Sci. Med.* Vol. 20, No. 2. pp. 153-166, Pergamon Press, UK.

Rice, Ronald E. and Frederick Williams (1987), *Theories Old and New: The Study of New Media. The New Media: Communication, Research and Technology*, Sage Publications, Beverly Hills.

Salzman, Harold (1989), 'Computer-Aided Design: Limitations in Automating Design and Drafting', in *IEEE Trans. on Engineering Management*, vol. 36, no. 4

Star, Susan Leigh (1989), 'Layered Space, formal representations and long-distance Control: the politics of information', *Fundamenta Scientiae*, Vol. 10, No. 2, pp. 125-155, Brazil.

Türk, Klaus (1989), *Neuere Entwicklungen in der Organisationsforschung. Ein Trend Report*, Stuttgart, Enke Verlag.

Weick, E. Karl (1990), 'Technology as Equivoque: Sense making in New Technologies', in Goodman/L.S. Sproull and Ass, *Technology and Organizations*, Jossey-Bass, San Francisco, Oxford pp. 1 - 44.

Wobbe, Werner (1986), *Menschen und Chips*, Sovec, Göttingen, p.10.

Zahrly, Jan (1990), 'Organization Design', in L.R. Gomez-Mejia/M. W. Lawless, *Organizational Issues in High Technology Management* ,Jai

Press Inc., p 79.

4 Developmental paths of work, technology and labour relations in Germany

H. Hirsch-Kreinsen

1 New technologies and the factory of the future

In the Federal Republic of Germany, the process of automation in the steel industries is characterized by an advancing diffusion of computer based and integrated systems of manufacturing (CIM). Many observers agree that new types of rationalization strategies and a change in modernization policies have emerged in industry since the 1980s. Stagnating economic growth and fiercer competition on both national and international markets call for higher levels of productivity and a more flexible production, if customers' demands are to be met.

The rapid pace of development in micro-electronics and information technology suggests that technical resources will reach a new level of optimization between high productivity and high flexibility, which have been contradictory goals up until now. Significant changes can be expected in more and more areas where the new computer aided and computer integrated manufacturing and control systems are employed. Due to the system character of these technologies, accompanying rationalization patterns are supposed to be both far-reaching and of a "systemic" nature. Computer integration offers a wide range of possibilities for organizing the factory of the future.

Future competitive factory structures are not only a result of technology and modern computer systems in production; according to practical experience, as well as in the opinion of many experts, questions of process organization and the design of work are of equal or even higher importance for realizing the potential for flexibility and productivity of new production technology. In the meantime it is widely agreed that the concept of the highly automated "unmanned" factory will find - at least in Europe - only very limited applications within the near future, and that this concept is, at best, suitable only for specific areas of highly standardized mass production. There is a growing body of opinion stressing the limitations of the long-prevailing Tayloristic or Fordistic model of manufacturing organization and work structuring (cf. e.g. Piore, Sabel 1984; Brödner 1985; Martin, Ulich, Warnecke 1987). This rationalization pattern is based on a distinct division of labour and job specialization on the shop-floor, and means centralized planning and decision-making. Tayloristic rationalization was particularly important for large industries and mass production during the 1960s and early 1970s in order to overcome manpower shortages in an expanding economy. This type of rationalization strategy has not only dominated in mass production industries in the Federal Republic of Germany, but has also been an important guide-line for mechanical engineering industries, characterized by customized manufacturing of complex products in small batches.

Essential preconditions for competitive factory structures in the future will not only consist of computer utilization and more and more computer integration in manufacturing, but also of skilled, qualified manpower, in these areas in particular. "New production concepts" (Kern, Schumann 1984), and new factory structures are being advocated which take advantage of the existing skills and qualifications and which seek to secure the potential for innovation and the abilities in adaptation, particularly in the case of the often smaller and medium-sized companies, by keeping the scope for taking action and decision making close to the basis of the manufacturing process - on the shop-floor. In the long run, it will only be possible to utilize the potential for flexibility and productivity offered by the new manufacturing technology through skilled workers. Skilled and cooperative work within production islands on the shop-floor is often mentioned in this context as an effective way to reintegrate and reorganize tasks and functions. This concept means a reduction of existing hierarchical organization patterns within companies, a decentralization of in-house decision making structures, and in general a reversal of the traditional "Tayloristic" rationalization strategies geared to distinct forms of division of labour and job specialization.

According to assumptions of this kind, some of which are of a decidedly prognostic nature, the development of industrial labour is currently undergoing a period of transition which, provided that adequate political support is granted, may easily lead on to new forms of industrial labour involving more broadly defined jobs and demanding less but skilled manpower.

It is not our intention to discuss what theoretical or empirical criticism might be brought forth on these assumptions.[1] Rather we would like to present a number of findings and hypotheses in the following, which will serve to draw a more differentiated picture of the present and foreseeable situation of industrial labour in Germany. Above all, it is important to avoid drawing the hasty conclusion that the implementation of new computer based and integrated technologies can be equated with an innovation of work organization.

Due to the widely diverging internal and external conditions there is, in our opinion, no uniform pattern currently emerging for the future development of industrial labour. On the contrary, the individual company has the choice of a broad range of options in technology as well as in work design. Presently we see a diverse picture of the development of industrial labour in West Germany.

The following argument is based on projects on technological and organizational change in West Germany industry conducted by the Institute for Social Research, ISF Munich. These research projects were concerned with the design potential with regard to work organization and skilled labour in the introduction of new technologies in the manufacturing area of the West German capital goods industry.[2]

It hardly has to be mentioned that the following findings and considerations about the development of industrial labour apply mainly to the situation in *West* Germany and to some degree to industrial developments in other Western market economies. After unification in 1990 it is still too early to judge their specific relevance for industrial companies out of the former G.D.R. For companies in the so-called "New Federal States" (die neuen Bundesländer), the transition to market economy is linked first and foremost with very fundamental restructuring problems. This process involves dissolving the "Kombinat" associations, redefining the production programmes, coping with financing and liquidity problems, creating competitive cost structures and establishing market-oriented distribution networks for the first time ever. Simultaneously, however, and with increasing urgency the question arises as to which *technological and organizational production structures* will be appropriate in this completely new situation and which measures should be taken to modernize production processes.

Western concepts on how technology can be introduced in an efficient and socially acceptable way and on how labour, equipment and materials can be optimally combined, provide a highly important orientation for modernization strategies in the NFL. At least some of these ideas are being taken up by management and consultants. At the same time, it can be assumed that these concepts cannot be transferred lock, stock and barrel to the former G.D.R. The problem we are facing is rather one of elaborating development plans and strategies which take into account the specific circumstances of the NFL-companies. First analyses in fact indicate that one of the greatest problems at the moment is that of preserving the innovative potential of former GDR economic units, which no doubt does exist, and putting it to purposeful use.

2 The capital goods industry in West Germany

As is known the capital goods industry plays a desisive role in the West German economy. In 1988 it employs approximately 3.8 million employees (about 18 % of all West German employees in the F.R.G.) and it has a turnover of about 713 billion DM (Statistisches Bundesamt 1989). The capital goods industry is a very heterogeneous part of the economy, including, among other branches, machine building, the automobile industry, shipbuilding, the electrical industry, office equipment manufacturers, and the aerospace industry.

Within this industry, machine building is the industrial branch which employs the most workers, followed by the electrical industry. These industries are superseded in turnover by the automobile industry which, however, takes third place in the number of workers employed. Thus, somewhat more than three quarters of all workers employed in the capital goods industry are employed in these three branches. The other branches follow at only a vastly inferior magnitude regarding both the number of employees and turnover.

In terms of product complexity, series size, and the dominant forms of production organization, the capital goods industry can be characterized as displaying a continuum in which the machine building industry represents the one extreme with high product complexity, small series and job shop manufacturing of discrete parts. The production of iron, sheet metal and other metal goods represents the other extreme with low product complexity, high volume mass production, and sequential line production.

With regard to job structure and work organization, two ideal types can be distinguished in the capital goods industry of West Germany today (c.f. Köhler, Schmierl 1992). These are skilled worker production and semi-skilled worker production:

(1) The job structures in systems of semi-skilled worker production are marked by a high degree of both functional and task specialization. The technical offices dealing with process and production planning, as well as the service departments for maintenance and quality assurance etc., play the decisive role in the planning, control, and supervision of the production process. The majority of production workers are mainly semi-skilled machine-operators whose skills are not only specialized in specific production techniques (such as drilling, turning and milling), but also in certain types of machines and a narrowly limited spectrum of parts. This horizontal specialization is supplemented by a clearly defined vertical division of labour, ranging from unskilled labourers to machine-operators of various levels of skill and experience, to relief men, job setters, group leaders, and foremen. The division of labour between job setters and operators is not clearly defined. In routine jobs, experienced operators may set up their machines themselves whereas more complicated tasks are carried out by job setters.

The work is organized in such a way that newly hired workers can be highly productive without comprehensive training. Paths of mobility from less skilled to more challenging jobs makes gradual on-the-job training possible, a process which takes place with little support in the form of instruction by superiors or classroom training.

(2) In systems of skilled worker production, the division of work according to functions and skill is less pronounced. The machine operators

take on small parts of the planning and service functions. Thus, to a certain extent they determine the job sequences on their machine, read process plans and decide about processing steps themselves, and can carry out less complicated maintenance tasks. Also the vertical and horizontal division of labour according to tasks/skills within the production departments is less pronounced, compared to systems of semi-skilled worker production. Skilled workers carry out all necessary production tasks themselves - as a rule including setting up complicated parts and machines. They are often capable of working with a variety of machines involving a number of different processing methods.

Skilled workers are normally trained within the scope of the so-called "dual system" of regulated, formal on-the-job training in firms or partially in external training centres and in part-time vocational schooling. More than 50 % of those in the 16-19 age bracket have such three to three-and-a-half year basic vocational training, for which the trainee, upon completion and examination, receives a certificate as a "Facharbeiter" or skilled worker.

Both systems of work are related to one another to a certain extent. They can often be found side by side in one enterprise which has different production processes according to series size and product complexity. Systems of skilled worker production dominate in customized, small batch production; systems of semi-skilled worker production dominate in mass line-production.

These systems of work are relatively new from a historical point of view. Their history can be described as a process of a growing differentiation of functions and tasks and an increasing narrowing down of skill profiles in production work. This is definitely true for systems of semi-skilled worker production which were generalized in West Germany after World War II, in the context of the dominating Tayloristic model of rationalization. It is also true for the systems employing predominantly skilled workers whose jobs, just a few decades ago, had included a wide variety of tasks in both the planning and service functions aswell as production tasks in the narrow sense, and at this time could be characterized as traditional forms of craft production.

3 On the development of technology and industrial labour

According to our survey findings[3] the application of CIM technology in the German capital goods industry is beginning to get under way in the second half of the eighties and is showing a considerable trend towards dynamic diffusion. In 1986/87 for example, only 9% of the companies in this industrial sector had already set up an in-plant, on-line integration between at least two functional areas of the company (such as design, process engineering, or production control departments). Information on planned computer integration suggest a trend towards rapid diffusion; almost a quarter of all companies in the capital goods industry are about to establish

more complex CIM systems and networks by the beginning of the nineties. Apart from larger companies, an increasing number of medium and small size firms are making steps in this direction.
Some of the company planning concepts involve extensive integration goals:

- About half of the companies already actively engaged in computer integration in 1986/87 planned to extend these facilities to encompass at least four different functional sectors within the next two or three years.

- Half of the companies planning systems were "newcomers" without any previous experience in computer integration.

- Approximately every tenth "newcomer" was about to take the major step towards a complete computer integration of their production process.

Even if some of these projects did contain certain "planning illusions" they certainly heralded a considerable and dynamic shift towards the computer integrated factory. Apart from the increasing utilization of flexible manufacturing systems and manufacturing cells the application of computer integrated organization and planning technology is one of the main fields in which CIM solutions will be realized. According to their point of departure and the rationalization objectives aimed for, the majority of companies are pursuing CIM strategies involving the realization of CAD/CAM or PPC systems (Production Planning and Control).

With regard to the design of work organization and workers' assignment there is currently not one predominant tendency to change to be seen. Our case studies, which were conducted on a broad basis in the mechanical engineering industry, reflect this situation by documenting the fact that these integrated systems will be operated to a large extent in a conservative way concerning forms of work organization and work structures.

In the course of system introduction about half of the companies investigated were found to present a considerable degree of stability and endurance in the existing forms of work organization and related skills structures, which are based on more or less distinct forms of division of labour and job specialization. Within the framework of this *"conservative strategy"* concerning work organization, the established division between conception and execution in company hierarchy is retained, as well as a marked division of labour with regard to functions and skills between process engineering departments, other services, and the shop-floor. In companies following this strategy the new systems are usually implemented by means of a short term oriented innovation process seeking to minimize risks of all kinds. Management planning, often with merely formal and passive participation of the works council, is almost completely oriented to criteria of technology, while design of work organization and manpower policy are regarded as being secondary problems. By proceeding in this manner the new systems are more or less thrust on to established company structures of technology and work organization without any further planning having taken place.

In only one third of the companies surveyed at least a partial departure from such policies has been observed, which one might term as an *"experimental strategy"* involving structural change in work organization. The main characteristic of this kind of strategy consists of experiments in changing traditional patterns of division of labour and company hierarchy.

According to this experimental strategy on the one hand, certain planning functions which were formerly performed outside of the workshop are transferred from the process engineering and planning departments back to the shop-floor, for example shop-floor scheduling or NC-programming. In those cases where mainly unskilled and semi-skilled workers had previously been employed, more jobs are now held by skilled craftsmen. However, on the other hand, an increase in the existing division of labour can also be observed - at times in different production lines within the same company.

Therefore the further course which those experimental strategies will take remains open in many cases. It cannot be assumed that a permanent and company wide reversal of the previous forms of division of labour will occur when such strategies are opted for. On the basis of a comparatively systematic innovation planning, which also considers questions dealing with organization of production processes and work, the management of these firms, in some cases together with the active participation of the works council, is engaged in a trial and error process, as far as technology and work organization are concerned, whereby the initial priority would seem to consist of keeping options open and not closing the door on future design potential.

As opposed to the two afore mentioned strategies, only roughly one tenth of the companies are pursuing a strategy of work organization which can be described as being *"innovative"*. The company wide departure from conventional "Tayloristic" strategies of designing work and organization in the implementation of CIM systems forms the essential attribute of this strategy. So far, in some of the companies investigated, this concept mainly appeared as a newly formulated "management philosophy"; in other cases a relatively extensive reduction of planning and engineering departments in favor of a re-integration of planning *and* operative functions into the shop-floor area was already under way - mostly based on the qualifications of skilled craft workers. Common to many of these companies is the fact that innovative strategies of work organization are pursued as the result of increasingly unfavorable conditions on the sales market. Such crisis situations arise when companies attempt to reconcile the need for flexibility on the one hand, and cost minimization on the other, or also when a very unfavorable cost structure is given due to high fixed costs, thus enhancing risks of running out of business. This also means that top management, as well as members of the representation of workers' interests pursue such innovation processes with a high degree of commitment, which is one of the preconditions for their success.

4 Three models to organize labour

Although the changes of work organization occurring with the implementation of computer integrated systems would generally not appear to be very dramatic, several associated factors exist which point towards a widely diverging range of developmental possibilities of future industrial labour. On the one hand these are aspects which are already evident in rudimentary form in the strategies of work organization outlined above, for example in the strategy involving structural change. On the other hand, due consideration must be given to the evident and often discussed changes of company external factors such as the structures of sales markets, labour market conditions or the further development of technology. Our research findings allow the identification, without any excessive stylization, of three ideal-typical developmental options of industrial labour:

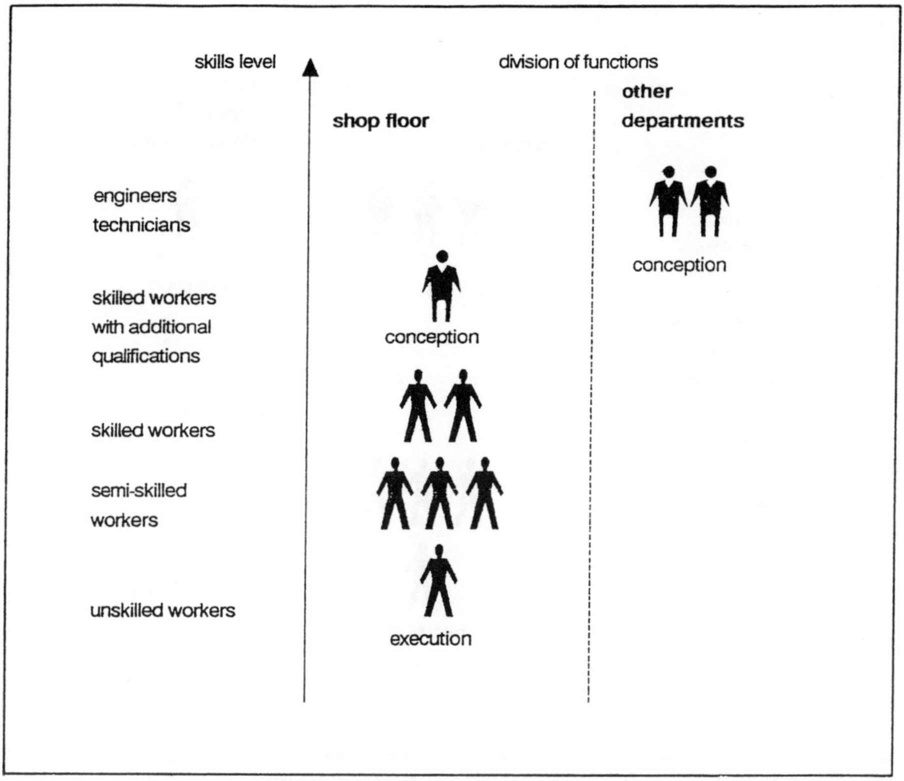

Figure 4.1 Computer integrated Neo-Taylorism

(1) Without doubt, the first developmental option can be characterized by the continuation or further consolidation and entrenchment of the traditional path of work design which can be termed as *"computer-aided Neo-Taylorism"* (Fig.1). Its central feature is the efficient enhancement of hierarchical, functional and skill related division of labour and the large scale use of computer-aided technologies. The functions of planning, programming or scheduling are exclusively performed by engineering departments. Service functions such as maintenance, repair, quality assurance and operational equipment management are systematically removed from the shop-floor and handled by specially trained skilled workers and technicians. The functions remaining on the shop-floor are those which are directly related to the actual manufacturing process such as machine setting, workpiece handling and machine monitoring. Even those mainly manual tasks are dealt with as far as possible by specialized workers in hierarchically graded jobs. The performance of these functions requires little more than rudimentary aptitude and the will to perform routine work.

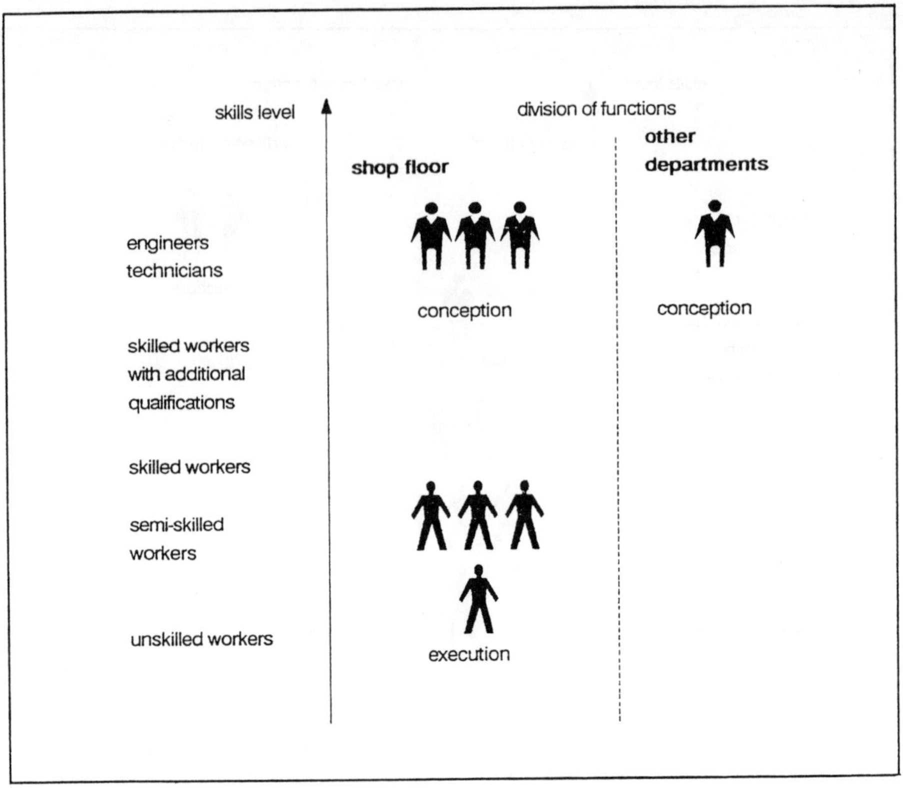

Figure 4.2 Skilled and cooperative production work

(2) In stark contrast, a second option is characterized by an extensive and lasting elimination of the traditional division of labour. This option can be defined as a concept of *"skilled and cooperative production work"* (Fig. 2). It is based on a considerable innovation of the existing structures of work organization and presupposes the existence of a workforce of qualified workers possessing skills on a relatively homogeneous level. According to this model, the workers are assigned with a whole host of inter-relating planning, monitoring and operative tasks and functions, ranging from process planning and programming to actual machine feeding and maintenance. The execution of these tasks is based on the principles of self-determination and cooperation. The concept of skilled and cooperative production work has been realized to the greatest extent in various forms of group work. Such work groups, like the system operators of a flexible manufacturing system, are to a large extent self-determined. Sometimes these groups will be integrated into the functional context of the manufacturing process according to the black-box principle, as "plant within a plant".

(3) By contrast there is a third developmental option which can be termed as

"polarized production work" (Fig. 3). Design strategies corresponding to this option make use of the obvious efficiency gains connected with even an only partial re-integration of planning, conception and execution into the actual front line of tasks performed on the shop-floor. According to this option, however, the planning jobs in the manufacturing area are not occupied by skilled workers or by workers having received further vocational training, but instead by personnel with an educational background more similar to that of technicians or engineers.

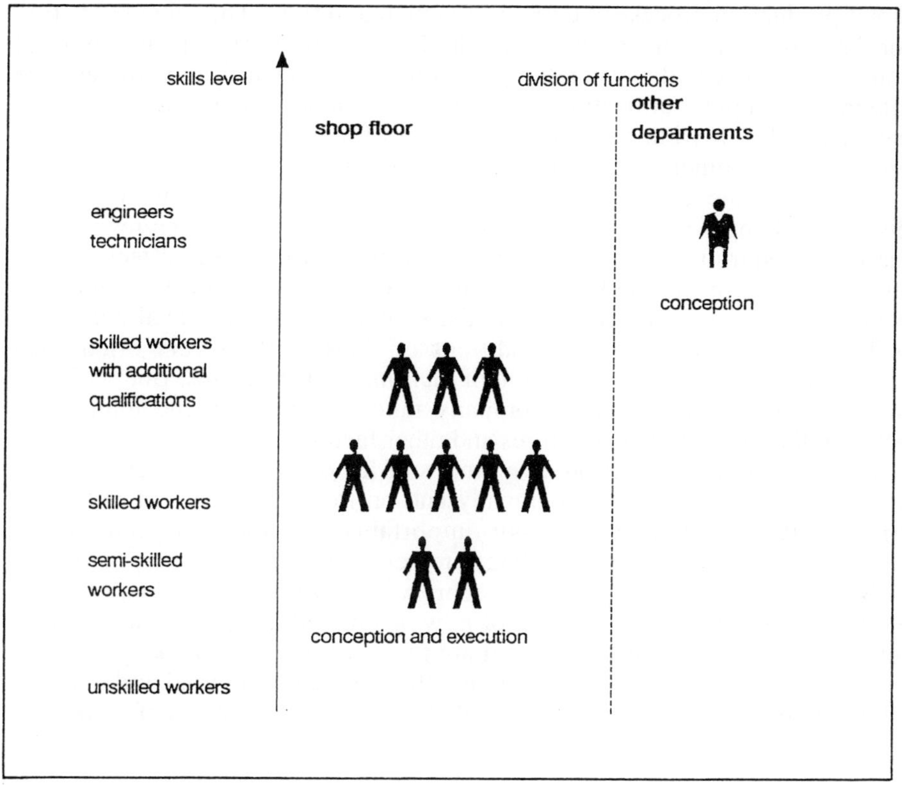

Figure 4.3 Polarized production work

On the basis of production systems designed accordingly, such members of staff hold new key positions in the manufacturing process - at central and complex shop-floor control centres managing the control and monitoring of entire manufacturing sections, for example, or supervising and controlling flexible manufacturing systems - positions which are endowed with comprehensive scope for decision making and taking autonomous action. The remaining functions to be dealt with on the shop-floor are again cut down to simple manual tasks, which can then be coped with by semi-skilled or even unskilled workers.

A comparison of these three options of industrial labour prompts the question what factors determine which of these options a company or a group of companies take. Obviously the company strategies of work organization are no longer influenced by factors which can be easily identified.

Firstly, work organization is no longer determined by technology. Progressive automation in manufacturing technology makes it possible to organize human work in many different forms and to disengage it from the direct production process in terms of function and time. The same holds true for data processing functions. Whilst in the past there was a relatively rigid connection between the generation, transformation and utilization of data, the integrated computer utilization across the limits of departments and functions has resulted in a wide range of possibilities for new combinations of data processing and functional and task specialization.

Secondly, due consideration must be given to the evident and often discussed changes of company external factors such as the structures of sales markets, labour market conditions or the further development of technology. But according to our findings there is no longer any uniform dependency between these structural factors and the strategy of an individual company. Without doubt the changing conditions of sales markets very often push companies towards a reorganization of their internal structures. But, which of the developmental options a company takes depends on the interaction between the sales market structures and other factors.

Concerning this interaction, two groups of factors can be percieved on the micro level of an individual company and on the macro level of industrial and social structures with major importance for the direction of the companies' strategies. These findings are generally somewhat surprising. As well as Industrial Sociology, labour economics have provided a multitude of evidence over the last thirty years of how macro and micro-economic forces (classical variables typically applied are the sales market, company size and series size, the level of automation, and the company's internal and external labour market) determine the structural development of systems of work.

5 The important role of power and policy

On the micro level of the individual company, factors of power and policy gain an increasing importance for the development of work organization and industrial labour. These factors are connected with the *implementation process* of computer based technologies. Undoubtedly, implementation processes play not only an important but also quite an autonomous role in determining the development of production and work organization. Management planning concepts, agents involved in the implementation process as well as resource planning particularly referring to know how, time and money available all seem to be the most significant.

So far, in many companies the introduction of new technologies is regarded as being primarily a problem of justification and technology, oriented to the selection of economical and appropriate hard and software solutions. The scope of company planning is restricted to short-term and limited considerations and the design of work organization does not come into consideration. In this context the interests and objectives of the managers and groups involved in the implentation process play a major role; normally they tend to stabilize the existing work structures. Within the implementation processes three types of managerial interests are of major importance:

(1) Top management tends to participate only periodically in the implementaion process, approving the investments and defining the global objectives, thereby determining the volume of investments, an approximate time table for the individual steps and phases of an innovation, and also the basic technical features a new manufacturing system. The profitability of a new technology is assessed with the help of an approximate calculation based on general economic indicators. In many cases only the direct system operating costs are taken into consideration, not the total costs which comprise direct and indirect costs. Overhead costs, such as the costs of the hierarchical division of labour within a company (the costs of the work planning department), remain excluded from the assesment of economic efficiency. Sometimes members of the top management participate in a general discussion about the future of work, but only very seldom do they use the results of this debate for their daily decisions, and for following innovative and new paths of rationalization strategies.

(2) The existing work structures are stabilized by the fact that the people who carry out implementation measures are members of the middle management. They develop an actual innovation concept within the global guide-lines which have been put forth by top management. As a dependent planning group, they are likely to be interested in pursuing strategies bearing the least risks and which would seem likely to guarantee operational efficiency within a short period of time. Furthermore, they can use the implementation of new technologies to strengthen or even expand their own position within the company. Their expert knowledge of technology and machine concepts is instrumental towards achieving this end. This means that as far as the question of work organization is concerned, system implementation rests on the established structures right from the start.

(3) The group of shop floor supervisors, production department supervisors and foremen traditionally take a sceptical stand on the issue of organizational changes. Formally speaking their position is a subordinate one with regards to managerial decision making, yet this group does exert considerable influence not only on the installation and initial operation of new manufacturing equipment, but also on the actual design of work organization and manpower assigment. Foremen and supervisors are relatively independent in their decisions concerning the staffing of jobs and the selection of workers. As is the case with middle management, foremen and

supervisors adhere to accustomed and established methods because their main interest is to maintain their leading positions on the shop floor.

On the one hand, this implies a rejection and by-passing of extensive Tayloristic forms of organization which would be connected with a loss of their authority and competency. Their interests are supported by many years of shop floor experience; they are aware of the fact that it is impossible to deal with all potential manufacturing problems by pre-planning and that it is the shop floor personnel who must come to terms with these problems in the end. This holds true for many sectors of the capital goods industry. On the other hand the prevailing structures of division of labour - and those which also establish themselves in manufacturing practice - are never basically jeopardized. The use of extensively trained workers to whom a certain amount of autonomy is formally conceded could also equally endanger the position of the foremen and supervisors. In this case the historical erosion of the position of this group would be increased.

Works councils do not normally intervene in this constellation of managerial interests in a big way. Instead, they merely play a subordinate role in the implementation process of new technologies. Not only because of their limited rights but also because of a lack of ressources and know-how, their policy is concerned with traditional fields such as wages, personnel policy or ergonomics. The works council's experience and knowledge was adequate for dealing with conventional kinds of changes, technology and organization. These changes effected limited and clearly defined cases and areas within a company. The works councils could draw on a wealth of experience they had gathered in dealing with the implentation of convetional machines and systems, and were able to foresee the probable effects on the employees and reduce the negative effects to a certain extent.

The introduction of new computer integrated technologies, however, does not simply imply that more changes will occur or that they will take place at a faster pace. The individual members of the work councils increasingly lack the necessary know how to even perceive the computer integrated rationalization measures taking place and their subsequent, far reaching effects on employees, as well as the new possibilities for work design. The experience the work councils have gathered so far is rapidly becoming obsolete.

The reslt of this situation is that whilst implementing new technologies work councils very often try only to keep existing jobs and wages, their differences and the differences connected with the employee's status. Actually this policy is more interested in preserving the existing organizational structures and the division of labour than in innovating the work organization.

Generally these interests are based on the company's existing organization with its more or less structured functional and hierachical division of labour. These overhead structures of the manufacturing process are eager to draw the newly arising complex of problems into their own areas of responsibility, i.e. to define them so that they appear to be genuine cases of application of their respective areas of competence. As an unintended result of this situation, companies are often pinned down to a conservative pattern of action, keeping the existing organizational structures.

6 The role of education and labour market

Within in the diverging factors on the macro level of industrial and social structures, we can see structures of major importance for the direction of the companies strategies for the labour market.[4] These conditions have a central mediating function in determining the size and skill structure of the workforce available for industrial work. Our assumption is that it will be increasingly difficult to recruit skilled workers for production jobs in West German companies. This diffuculty is particularly an obstacle against the broad diffusion of "new production concepts" and a "reprofessionalization" of industrial labour in general (Kern and Schumann). Such concepts as well as the option of "skilled and cooperative production work" necessarily presupposes that enterprises can recruit and employ a particular type of worker fairly easily, for which the German industrial skilled tradesman can serve as the prototype, and can keep him long-term. Workers of this type must combine considerable intelligence with the willingness to perform physical, at least at times, monotonous and mentally demanding (and for this reason not particularly high status) work; he must be willing and able to gather experience in practical work over a long period of time and to supplement this experience with ever new technical knowledge.

In most industrial nations before the historic constellation of the "Tayloristic syndrome", a workforce of this type in more or less large numbers was available virtually as a matter of course. Since their origins, heavy industrial enterprises found a stable and reliable reservoir for procuring skilled labour or workers who could be easily trained to perform production work in the social and economic structures of the traditional agricultural sector. Thus, the industrial skilled tradesman - including the traditionally and still very significant portion of technical personnel in the German-speaking countries who have been promoted from the skilled worker ranks - is embedded very deeply in the social structures and cultural traditions of the pre-industrial "people". It is here one finds workers willing to perform physical, often very taxing and not particularly well-paid work, and at the same time accept responsibility for people, machines and material being dealt with, and who are able to learn various techniques, put them into practice and develop them. This phenomenon is only imaginable against the background of conditions in which hard physical labour was the inevitable fate for children in this social sector and skilled production work in industry was regarded as a real opportunity.

The coming of the Tayloristic-Fordistic pattern of rationalization and economic growth brought the total erosion of the conditions which ensured industry's supply of intelligent and competent production workers. Two

processes interacted in this development:
(1) Firstly, the highly efficient mechanisms with which manpower for industrial wage labour was mobilized in former times, slowly but surely dried up the social and economic structures in which the manpower had grown up and been socialized in traditional rural living conditions and production modes. Even in the German speaking cultures, which were able to keep the tradition of industrial skilled work alive longer than other countries, virtually no children who are normally gifted would consider industrial work as an acceptable prospect under the conditions largely applicable in industry today. And, massive immigration from less developed countries has been receiving more and more resistance over a long period of time.
(2) Secondly, through the generalization of behavior oriented to utility maximization prevalent in modern industrial nations, the human capital calculus has become the dominant factor in selected educational and occupational paths. The massive rush for high level education (towards university or quasi-university education) unleashed by these trends, has made schools, forced to engage in more and more scrutinous selection procedures, the main means of distributing professional and social chances in just a few decades; in the short or long-term, graduation from institutions of higher education is going to be the necessary prerequisite for access to all the attractive and better paid jobs. Furthermore, in the course of a generation, the traditional forms of social reproduction of the skilled or highly skilled industrial workforce will lose all meaning through practical training and practical experience.

To the extent that these two processes converge and are not counteracted by political measures, which would require both great courage as well as exceptionally favorable circumstances, enterprises will find intelligent employees with utilizable industrial skills and knowledge exclusively among the graduates of higher semi-academic or academic institutions more and more frequently.

Under these conditions, companies wishing to overcome the inefficiencies of a structure characterized by a marked division of labour and to avail themselves of the advantages of an organization based on autonomous units of production, capable of reacting flexibly and innovatively, they must create the necessary conditions (in work organization, career prospects, payment, etc.) to attract graduates from institutes of higher education. Work organization systems following the principle of "reprofessionalization" of production work and a far reaching reduction of division of labour probably do not to fit these conditions.

7 The future of industrial labour and the task of future research

Conclusions that can be drawn from our findings are that extensive and rapid changes in technology and labour have taken placen under the influence of various factors, such as the international division of labour, changing consumer behavior, and the availability of new technological components and systems. It seems to be clear that the observed processes of adjustment are leading in various directions. However, they have different chances of being put to practice:

(1) The "computer-aided Taylorism" should assert itself without doubt. It will certainly be connected to a loss of performance, efficiency, and profitability due to the fact that turbulence in the sales markets and the associated flexibility requirements are on the increase. On the other hand, one must also give due consideration to the computer based capacity of Neo-Tayloristic structures in order to neutralize the objective pressure towards change over a longer period of time successfully.

(2) While there are good reasons to assume that only industrial work organized according to skill based concepts will be capable of guaranteeing long term economical efficiency, particularly in view of the increasing flexibility requirements stemming from the sales markets, it is doubtful as far as the present and foreseeable conditions are concerned, that this form of work will assert itself as the main course of development. Firstly obstacles based on existing organizational structures of companies are not to be overlooked: the risks involved in departing from well established organizational structures and the associated shift from previous power and interest constellations. Secondly obstacles concerning the decreasing supply of skilled and qualified production workers on the German labour market should be taken into consideration.

(3) Particularly in comparison with the skill based concept, the option of "polarized production work" would seem to have considerable chances of being put to practice. There are considerable advantages to be gained with regard to flexibility and efficiency by a partial reorganization of company division of labour, while avoiding the risks of leaving evolved and established internal structures and the corresponding constellations of power and interests. Although this option is highly problematic in terms of social concerns and social policies, it is probably an attractive developmental perspective for many companies. Moreover, many companies see few risks of finding the required personnel for the comparatively low number of key workshop related positions.

But, for the time being, it is not possible to draw a definitive picture of the future development of industrial labour in West Germany. This means that controversial positions in the debate on the future of industrial labour are not based on academic speculation, but rather that these positions can refer to tangible - albeit selectively interpreted - empirical evidence at any time.

This situation is not merely the result of not enough observation. A much more intensive description of current adjustment processes in technology and work and particular company policies would not bring greater clarity. The diversity of the observations and expected trends must be understood as a fundamental characteristic of the historical situation in which the advanced industrial nations presently find themselves. As already indicated, this situation is grounded in the erosion of the "Taylor syndrome" characteristic of the post-war period (Lutz 1992). This syndrome was determined by a set of macro and micro-economic forces such as the expansion of sales markets, economies of scale, standardization and "Detroit-automation" of production, and a corresponding technology market as well as a large labour force reserve. Tayloristic strategies of work design were an expression, as well as a component, of this constellation. However, this constellation seems to have dissolved during the 1970's and 80's. Micro and macro economic forces which were moving together in the past, developed into a multidimensional force field, whose vectors are directed in various and partially contractory directions. A new equilibrium is not yet in sight. The current disunity in the development of rationalization processes can only be understood in this context.

Consequently, according to Lutz (1992) as well as other authors (e.g. Piore and Sabel 1984 or Boyer 1989), one can maintain that realistic assertions about the future of industrial work can only be the starting point of research. The main task has to be to combine the developmental trends on the micro level of the individual company with the macro-conditions of technology and work systematically , and to analyse the historical processes that have been generated and changed by the interelationship of micro and macro-conditions.

Notes

1. For a critical discussion of the current debates on the future of industrial labor in West Germany see Bergmann (1988) and Altmann et al. (1992).
2. The conceptual background was developed in the project "B2" of the special research unit SFB 333 on post-Tayloristic rationalization strategies. The empirical findings were provided by survey and case study research carried out in the years 1985 to 1990 (see Schultz-Wild et al. 1989; Hirsch-Kreinsen et al. 1990). Members of the project group apart from the author, are Marhild von Behr, Burkart Lutz, Christoph Köhler, Klaus Schmierl, and Rainer Schultz-Wild.
3. These findings are based on 1,096 questionnaires filled out by West German companies in the capital goods industry. The results of the questionnaires were weighted by size of company and industry branch according to their proportion of employment in the West German capital goods industry. The survey was conducted during the autumn and winter of 1986/87. For more details see Schultz-Wild et al. 1989 and Schultz-Wild 1989.
4. The following arguments refer especially to Lutz (1990 and 1992).

References

Altmann, N., Köhler, Ch., Meil, P. (eds.) (1992), *Technology and Work in*

German Industry, London .

Altmann, N. et al. (1992), 'No End in Sight - Current Debates on the Future of Industrial Production Work', in Altmann et al. (eds.), *Technology and Work in German Industry*, London .

Bergmann, J. (1988), 'Technique et Travail: Bilance Critique des Recherches Allemandes en Socoilogie du Travail', in Cohendet et al. (eds.), *L'Après - Taylorisme. Nouvelles Formes de Rationalisation dans l`Enterprise en France et en Allemagne,* Paris .

Boyer, R. (1989), *Regulation Theory - A Critical Assessment*, Columbia University Press, New York.

Brödner, P. (1985), *Fabrik 2000 - Alternative Entwicklungspfade in die Zukunft der Fabrik*, Berlin.

Cohendet, P., Hollard, M., Malsch, Th., Veltz, P.(eds.) (1988), *L'Après - Taylorisme. Nouvelles Formes de Rationalisation dans l`Enterprise en France et en Allemagne*, Paris.

Hirsch-Kreinsen, H.; Schultz-Wild, R., Köhler, Ch., Behr, M. (1990), *Einstieg in die rechnerintegrierte Produktion - Alternative Entwicklungspfade der Industriearbeit im Maschinenbau*, Campus, Frankfurt/New York.

Kern, H., Schumann, M. (1984), *Das Ende der Arbeitsteilung ? Rationalisierung in der industriellen Produktion*, München.

Köhler, Chr., Schmierl, K. (1992), 'Technological Innovation - Organizational Conservatism ?', in Altmann, N., Köhler, Ch., Meil, P. (eds.), *Technology and Work in German Industry*, London.

Lutz, B. (1990), *Le Mirage de la Croissance Marchande - Essai de réinterprétation du développment du capitalisme industriel dans L`Europe du XXe siècle*, Paris.

Lutz, B. (1992), 'The Contradictions of Post-Tayloristic Rationalization and the Uncertain Future of Industrial Work', in Altmann et al. (eds.), *Technology and Work in German Industry*, London.

Martin, T.,Ulich, E., Warnecke, H.J. (1987), *Appropriate Automation for Flexible Manufacturing. Preprints of the 10th IFAC World Congress. VDI/VDE-Gesellschaft für Meß- und Automatisierungstechnik*, Düsseldorf, pp. 291-305.

Piore, M.J., Sabel, Ch.F. (1984), *The Second Industrial Divide - Possibilities for Prosperity*, Basic Books, New York.

Schultz-Wild, R., Nuber, Ch., Rehberg, F., Schmierl, K. (1989), *An der Schwelle zu CIM - Strategien, Verbreitung, Auswirkungen*, RKW-Verlag, Eschborn, Verlag TÜV-Rheinland, Köln.

Schultz-Wild, R. (1989), 'On the Threshold of Computer Integrated Manufacturing - Diffusion Trends of CIM-Technologies in West German Industries', in *CIMS (Computer-Integrated Manufacturing Systems)*, Vol. 2, No. 4, 1989, pp. 240-248.

Statistisches Bundesamt Wiesbaden (Hrsg.): Statistisches Jahrbuch 1989 für die Bundesrepublik Deutschland, Stuttgart 1989.

5 New technology and the working conditions of the print media

R. Hummel

Within the last decade, the media field of new technologies in particular has paved the way to an important change. There is a large number of publications dealing with "technology assessment" in the communication sciences, particulary concerning the introduction of the so-called "new media". The goal of this approach is more or less a "systematic study of the effects on society which may occur when a technology is introduced, extended, or modified, with special emphasis on the unintentional, indirect, or delayed impacts" (SLACK 1984, 15).

Furthermore, much empirical research has been carried out reflecting some special problems of working conditions of journalist in the "classic" print media caused by the usage of new technical tools or the disappearance of skilled graphical workers within these enterprises. Relative studies which are well known are mainly those which deal with word processing units and computer aided pagemaking in newspaper publishing, when journalists are no longer only responsible for the "intellectual" but also for the "technical" side of production. Nevertheless the installation of such "Integrated Systems" is not at all a question of technical feasibility: The first system was installed in 1963 in the USA, but in Austria, newspapers first began to be published in this way in the eighties.

However, there seems to be a lack of studies reflecting the totality of the people involved in the production process of newspapers (i.e. all kind of employees plus the employers), the totality of the production process itself (i.e. the differentiation between the parameters of change immanent to technology and independent of technology), and finally, the result of all these production efforts, the quality of newspaper itself. Technology assessment has to take into account not only the changing working conditions caused by computerisation, but also a historical shift in the way production makes possible the introduction of new technologies (c.f. WEIZENBAUM 1987). Building hypotheses stating that there is a change in working conditions presupposes the questions which are analyse the exact nature of the relative task, and what significances these tasks have for each person involved.

Therefore we decided[1] to follow a method described by Paul Felix Lazarsfeld (1941, 12) under the notion of "critical research" long before the label "technology assessment" emerged:

> *"a) A theory about the prevailing trends ... is introduced on the basis of general observations... b) A special study of any phenomenon consists in determining how it expresses these prevailing trends... c) The consequence of (b) in stamping human personalities in a modern industrial society are brought to the foreground and scrutinised from the viewpoint of more or less explicit ideas of what endangers and what preserves the dignity, freedom and cultural values of human beings. d) Remedial possibilities, if any, are considered".*"

Methodological framework

A sample of journalists, clerks, graphical workers as well as publishing house managers of the 28 biggest Austrian daily and weekly publications was interviewed in a methodological combination of quantitative and qualitative items. Such a proceedure enalbed thus the creation of a picture of some strategic points in the objective and subjective material, or, in other words, a statistical and clinical treatment of the outcomes (LAZARSFELD 1978, 15). Key questions concerning the introduction of computer aided newspaper production were:

- Changes of the amount of necessary work;
- changes of the qualification of employees (skills, training, capacity, relevence of prior qualifications);
- possibility of the workers to comprehend the entire production process (alienation, communication structures);
- changes within the special field of work (job enlargement or job enrichment;
- changes of the quality of products, the costs of production and the effects of competition;
- attitudes concerning the working situation of the respondents, technique in general, their belief in the future development and to the propositions of trade unions;
- further strategies of the concerned enterprises;

- summaries und conclusions to assess possible perspectives within this field.

The research framework consisted of four major parts: Firstly, managers und employees were interviewed on the afore mentioned topics. Secondly, an analysis of the working day.was integrated into the questionnaire. In total 79 different occupations during the working day of either a journalist, a clerk or a graphical worker, including "recreational activities" like coffee breaks or short chats with collegues was presented on a set of cards. In the third place, an output analysis of printing errors was made by professional proof-readers in a sample of one quality paper before and after the introduction of desktop systems. The two-step evaluation finally consisted of the commentaries of 20 journalists, graphical workers and clerks on the draft-interpretation and in a group discussions of their different contributions. Thus we wre able to incorporate the views of experts. The aim of this methological approach was not to gain more or detailed data, but to validate the interpretation and to find more explicative arguments.

Outcomes

The main reason for publishers to introduce new technology is to minimize the costs of newspaper production. Change of quality (be it graphical appearance or news gathering) do not play an essential role within this context. On average, more output is produced by less people as a direct consequence of computer aided publishing. Therefore this technology is introduced first of all as an investment for rationalisation, but not merely as a product improvement and also not for expansion.[2] (Nevertheless this technology would be excellently adequate to achieve such goals.) Following the prognosis of newspaper managers, there will be again a dramatical reduction in the number of graphical workers, a slight diminuation of clerks and an increasing number of journalists in the paper of the future. This is why the "informatisation" of newspapers consists not like the introduction of a given well-defined technology (as the linotype-setting was for example), but it is the start of an open ended process for the future. The next steps well be fully electronic pagemaking on the screen by journalists and computer to plate.[3] Further developments like teledistribution of newspapers via fax or the mutation of publishing houses into providers of news merely in electronic form are indeed technologically feasible but utopian, at least for middle term realisation.

The majority of employees asked in this research project, nevertheless have a positive attitude towards the further developments in the media field. The reasons for this lie in the improvement of control over the entire production process by a single worker due to the systematic structure of this technology. Moreover, the computerisation goes hand in hand with a real or presumed improvement of qualification, knowledge and responsibility of employees. Another reason for positive attitudes might consist in a strong personal identfication with the profession, especially in journalists and graphical workers.[4] Qualitative evaluations gave us strong evidence that expressions of general discontent with working conditions would not be made within a standardized interview to avoid professional role conflicts: Since journalism but also graphical work is treated merely as a profession, one has to be first of all talented, those complaining about the work situation could admit not to be gifted enough for this specific kind of work. Nonetheless, this generally positive view is contrasted by a series of single negative experiences, also verified by the working day analysis: journalists have to accept a shifting of their daily working hours (on average by 36 minutes!); graphical workers have to face a reduction in the diversity of labour; all professional groups are confronted with the intensification of their respective work.

Therefore, not only in the research area, particularly when dealing with so-called new technology, but also in our methodological approach, we have to think in a structuralistic way. Not only does the pure emergence of statements provoqued by interviews show the reality, but the understanding of different relationships in which the interviewed persons are involved (cf. BOURDIEU 1974). – In a way, this is a critique on prevailing trends in technology assessment in the field of media.

This specific case also shows the "border crossing potential" (BECK 1986, 348) of electronic systems – maybe the most important consequence: As the working day analysis proves, the specific duties of journalists, graphical workers and clerks unify more and more. All the three groups take over tasks from the other professions. Journalists are now also responsible for type composition, proof-reading, in some cases for correcting, layout and even some administrative work. Graphical workers become "technical editors" with competences in rewriting and research work – as do more and more clerks, especially those from editorial offices and advertising departments.

Figure 1 shows that according to our data journalists under conditions of "integrated newspaper production" spend about 75% of their working day in their genuine business, but the other two groups only approximatly 60%. This is first of all due to a take over of journalistic tasks by clerks and graphical workers.

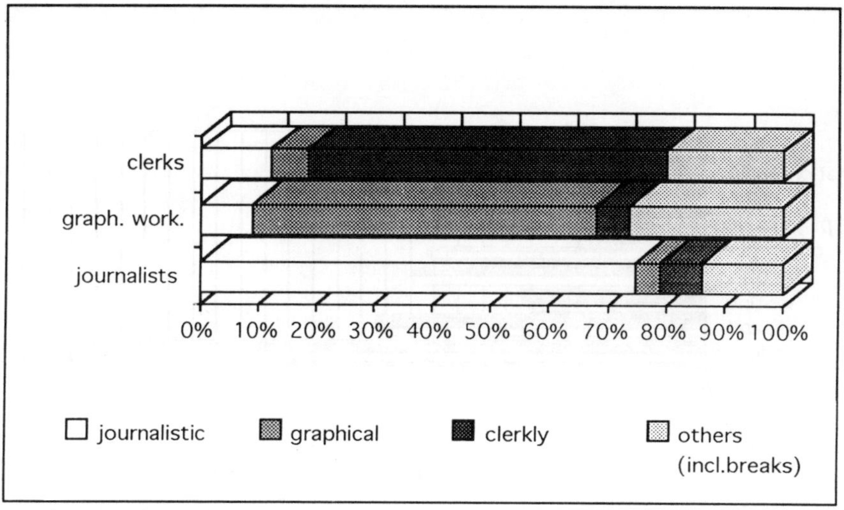

Figure 5.1. Percentage of occupational types during an average working day

The journalists responsibility for typography results in an increasing number of printing errors when professional proof-readers have been eliminated (six times more in the observed sample). The opinions of newspaper managers towards this phanomenon, expressed in the interviews, demonstrate two opposed comprehensions of this problem: Some are trying to keep proof-reading by graphical specialists also in the area of computerized newspaper production as was the case previously. Others put up with deteroriation of quality, being optimistic stating that consumers would not be annoyed by this fact.

Luckily enough not every department of the publishing houses in our sample was completely involved in new technology.[4] Therefore we had a control group not working with computer aided production. Figure 2 shows that journalists working with new technology spend a longer period of their working day not only in proofing and layout but also in editorial conferences and organisation than their collegues working in the "old-fashioned" way. This seems to be more evidence pointing towards the fact that new technology was introduced by numerous publishing houses in a more or less chaotic way following "trial and error".

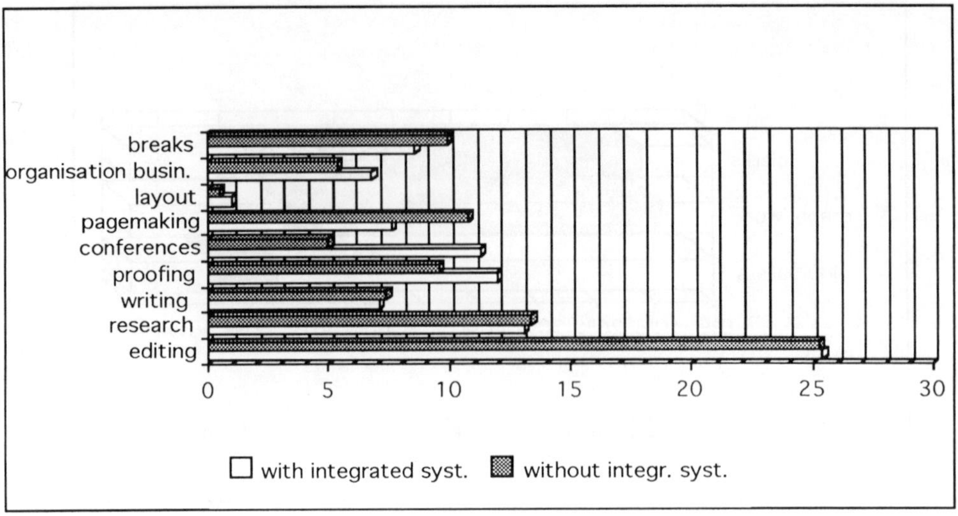

Figure 5.2. Differences between conventional and computer aided journalism (percentage of daily working time)

Both, managers and workers, showed a restricted knowledge of the technical standards and their consequences for strategic use. Computer-aided newspaper production was treated just as a new tool for every-day work by the majority of interviewed persons. So, the greater amount of working hours and stress for journalists is not a technical problem but a question of organisation. Inefficient organisational structures – and a much too small number of journalists – will not be corrected by computer-aided systems but on the contrary deterioriated. The big formative power inherent in this technology cannot be used if treated only as measure of rationalisation, or, as just another type of conventional machinery. On the other hand it is true, that hard- and software-development was not determined by the needs of publishing houses in many cases, but more by computer houses, since newspaper-computer-systems represent only a small segment of the entire computer market. – But the proposal to establish more power for the consumers (i.e. the newspaper houses) by cooperation in demand (ARBEITSLIVSCENTRUM 1985, 43) remains inefficient as long as the information level about technology stays low.

Conclusions

The computerisation of print media will accelerate the transformation of media professions, but the reason for it is not the technology itself. It is the organisational structure of work within the newspaper houses – this again a result of the self defined role of the different media and their market position

– which causes the concrete application of new technology.

The journalistic profession no longer consists only of news gathering and news interpretation, but also of technical and sometimes even of administrative work within the media production. Graphical workers will be either totally excluded from the pre-print sector or become some kind of journalists. Parallel to this, clerks cover more and more technical and journalistic functions, especially in smaller papers, because labour costs are much lower than those of the other newspaper employees. As a result a deterioriation of the professional status of journalists may be expected. This does not necessarily mean a dequalification in total but might lead to a less intellectual image of journalism in general (like printers made a decline from a kind of artisans in the Gutenberg epoque to the skilled workers of nowadays).

However it must be emphasised once again that new technology is not the intrinsic reason for this shift in professional roles, but it makes the change easier and will accelerate it.

Computerisation in the newspaper field causes two paradoxa:

- Firstly concerning the organisation of labour: The concerned professional groups are subjected to tendencies of integration and to taylorisation at the same time. In general, journalists, clerks and graphical workers will have more and more similar jobs but, on the other hand, a diversification within the branch between more creative and more administrative spheres. Quite similar stress for the users of Integrated Newspaper Systems is increased and typographic quality is deterioriated while at the same time computer-aided publishing reduces the amount of work necessary and gives more control over the entire product to the skilled pagemaker.

- Secondly concerning the direct effects: They are both spectacular and harmless at the same time. Integrated Newspaper Systems are a completely new way of print production and the often quoted allusion to the Gutenberg invention is less pathetic than it seems. But their introduction is in every case part of a process (not least because their development is not finished yet). In conjunction with the professional roles of the media employees, taking tribute to a long tradition,[5] the change in the working conditions is less visible – but it is nevertheless going on. In combination with the widespread miscalculation that information technology was nothing but a "normal modernisation", employers and employees will tend to accept "natural constraints" seemingly caused by the technology instead of developing strategies.

A crucial point for a well operating newspaper system lies therefore in consciousness about the aims of the production process, in knowledge about the technology in general and in the training of the use of system components in particular. Without special education in this field, such excellent poposals like participation of empoyees in decision making will remain purely academic.

As far as technology assessment is concerned (not only) in the media field, there is a indispensable necessity to embed the research in general sociological observations. Professional standards and traditions move far less than technical standards. Thus changing working conditions can not be conceived merely by analysing the immediate consequences of technology.

Notes

1. It is evident that every assertion here is made from the background of the Austrian situation.
2. This means the processing of printing plates from the desktop computer without printing forms.
3. Strong evidence for this could be seen from two studies I carried out in 1992 and 1993 about professional behaviour and personal values of journalists and graphical workers in Austria (Hummel 1993).
4. One reason for it lies in the agreement between trade unions and enterpreneurs concerning the so-called Integrated Systems in publishing houses: It is statued herein that every journalist has the right to refuse the work with such systems for a period of 8 years from date of the first installation in a given publishing house.
5. In Austria, as we mentioned this, affects not only journalists but also graphical workers. Their number was considerably reduced but in contrary to Great Britain they cannot be replaced all together by unskilled workers by reason of labour legislation.

References

Arbeitslivscentrum (ed.) (1985), *The UTOPIA Project. An Alternative in Text and Images*, Stockholm

Beck, Ulrich (1986), *Risikogesellschaft. Auf dem Weg in eine andere Moderne*, Frankfurt/Main

Bourdieu, Pierre (1974), *Zur Soziologie der symbolischen Formen*, Frankfurt/Main

Hummel, Roman (1990), *Die Computerisierung des Zeitungmachens*, Wien

Hummel, Roman (1993): 'Wir sind ganz normaler Durchschnitt, nur etwas besser...', Bericht über das GBV-Forschungsprojekt *Graphische Arbeiter 1993*, Wien

Lazarsfeld, Paul Felix (1941), 'Administrative und Critical Research', in *Zeitschrift für Sozialforschung*, vol. 9, pp 2 - 16, New York

Lazarsfeld, Paul Felix et al. (1978), *Die Arbeitslosen von Marienthal*, Frankfurt/Main

Sandberg, Ake (1984), *Alternative systems for text and image processing in newspaper production – some experiences from UTOPIA project*, Stockholm

Slack, Jennifer Daryl (1984), *Communication Technologies and Society*, Norwood/New Jersey

Weizenbaum, Joseph (1987), *Die Macht der Computer und die Ohnmacht der Vernunft*, Frankfurt/Main

6 Information technology in trade unions. Implications for decision making in trade unions

G. Stewart, L. Hamilton, S. Cockerill and J. Douglas

Introduction

Studies of the implementation of new technologies in general and information technology in particular tend to focus either on the "impacts", positive and negative, of the technology on workers and their jobs (i.e. on employment / unemployment, de-skilling / reskilling, job satisfaction / dissatisfaction, health etc.) or on specifying the environmental contexts of organisations most conducive to the successful implementation of technology based managerial strategies. The approaches vary from being "technologically deterministic" where the technology is treated as an independent variable, fixed and reified (Woodward), "economically deterministic" and used by management as a tool of economic and political domination over a recalcitrant labour force (Braverman) or "neutral", where the technology is introduced as a matter of "strategic choice" between more or less competing/co-operating interest groups within the organisation (Child).

In most cases the focus is on capitalistic or, at least, non democratic forms of organisation where the deployment and exploitation of the technology is purposive and has the intent of achieving organisational goals either in direct form of the production and distribution of goods and services for markets or in an indirect form of facilitating and supporting the internal administrative functions of the organisation.

Rarely is consideration given to an examination of the purposes, process and problems involved in introducing new technologies into non-capitalist, voluntary organisations where the goals of the organisation are, in theory, determined by the members of the organisations and where the leaders of the organisation are, again in theory, subject to the democratic control of the members. Yet, the capacity for the new technologies to alter fundamentally both the boundaries between the technical and social relations within such organisations is surely considerable.

The prime focus of this paper is the introduction of information technology into a trade union, the British National Union of Mineworkers (NUM). The focus is less upon the particular attributes of the technology itself than on the processes by which it was introduced. This involves examining two interrelated issues. One is peculiar to the NUM where the focus is on the extent to which such processes were consistent with the democratic and participative decision making processes of the union. The second follows from the first and seeks to examine more generally the potential for such technologies to enhance or erode the democratic and participative mechanisms within trade unions. Little research has been done in this area, the study by Rees and Owen (1993; see also Hastings, 1983) on the effects on democracy of computerising membership records in two unions being an exception though the lack of any concrete conclusions is somewhat disappointing.

The paper thus has a specific and peculiar focus. The objective is not, as one could reasonably expect, to attempt an empirical testing of hypothesised relationships within organisations such as trade unions. In fact there was no "research" as such. Rather, the stimulus for the paper arose as a consequence of one of its authors acting in a consultative capacity for the NUM. In the context of what might appropriately be seen as a "social action research" framework, his participant observer role led him to consider the nature of particular levels within the union hierarchy, their role in the introduction of the new technology and the potential for the latter to facilitate changes in the nature of the relations and processes within that hierarchy. In conjunction with the other authors the attempt was then made to find an appropriate theoretical framework within which to understand, explain and articulate the processes at work.

The paper has four substantive sections. The first examines the nature of the formal democratic mechanisms within trade unions. This section constitutes the theoretical background for understanding the process of trade union government and the extent to which information technology might facilitate changes in such.

The second section focuses specifically on the NUM, the peculiar economic and political context of this trade union and the nature of the participative mechanisms which operate therein. This acts as a precursor to section three where a case study of the processes of introducing information technology into the NUM is undertaken.

The final section is distinctive and important not so much for its role in concluding that particular processes are at work in the NUM or in trade unions in general. In fact conclusions are not drawn. Rather, the question is posed as to whether the alleged ability of information technology to facilitate changes in organisation‘s administrative and control structures applies equally to trade unions and, if so, what kind of research agenda might be constructed to analyse the issues involved.

1 Theories of trade union government

A full discussion of the democratic decision making processes in trade unions and the extent to which the introduction of information technology (IT) affects these, requires a number of areas to be considered. Inter alia, these would include an examination of the very nature of trade unions as organisations; definitions of and analyses of the distribution of power within such organisations; and definitions of the very concept of democracy in the context of these organisations. Since it is part of the purpose of this paper to stimulate a debate on what might be termed a "research agenda" and since these areas effectively constitute what the authors consider to be a major part of the problematic, these are not entered into here.

For the moment we wish simply to adopt the working definition proposed by Salamon that trade unions are organisations "whose membership consists of employees, which seeks to organise and represent their interests in both the workplace and society and, in particular, seeks to regulate their employment relationship through the direct process of collective bargaining with management" (Industrial Relations - Theory and Practice, 1987, p58).

We wish also to proceed on the assumption that since such organisations are voluntary, the decision making processes ought to allow for extensive membership participation in and determination of both the rules governing the organisation and the means by which its objectives are attained.

This allows us to focus more closely on the nature of the democratic processes themselves, the extent to which they allow for effective membership participation and how the introduction of information technology affects these.

Analyses of trade union governance have a long history. As early as the turn of the century writers such as the Webbs (1898) and Michels (1962; a work first published in 1911) investigated the extent to which trade unions could operate as "democratic" organisations as they grew. For the Webbs, the concepts of "primitive" and "representative" democracy were important in describing the changes in trade union organisation. In brief, the former described a situation and time when trade unions were small enough to permit the members to meet together, collectively make the rules and policies governing the organisation and decide how such might be applied. Increased membership rendered such processes inoperable and attempts to retain the qualities of "primitiveness", resulted in "either inefficiency and disintegration or to the uncontrolled dominance of a personal dictator". Gradually, therefore, unions underwent a "silent revolution" whereby, in place of the primitive assembly, there emerged the "typically modern form of democracy, the elected representative assembly, appointing and controlling an executive committee under whose direction permanent official staff does its work" (S. and B. Webb, 1898).

Coincidentally with the transformation from primitive to representative democracy, and concomitant with increasing size was the replacement of an "amateur", elected administration made up of ordinary members, by a "professional", sometimes elected, sometimes appointed administration made up of people with skills acquired through years of education and experience in administration, skills which the rank and file neither possessed nor could acquire. As Banks puts it, "the appointment of a General Secretary and his gradual reinforcement by a staff of regional and divisional organisers, research officers, and other office personnel were in one sense no more than the increasing division of labour which seems to result inevitably from an expansion in the scale of operations". (Banks, p87).

Banks also suggests, however, that this could lead to problems of democratic control and this is the major theme of Michel's observation that there exists in labour movements an "iron law of oligarchy" whereby a leadership clique ("oligarchy") eventually becomes immune to the democratic processes. A major element in this "law" and one pertinent in the current context is Michel's reference to the technical competence of leaders which elevates them above the mass and subjects the mass to them. "This special competence which the leader acquires in matters inaccessible , or almost inaccessible, to the mass gives him a security of tenure which conflicts with the essential principles of democracy". (Michels, p109-110).

The outcome of these processes then is centralised decision making by the oligarchy. Centralising tendencies may be counteracted by decentralising tendencies resulting from the actions of a minority of disaffected leaders who, bound by national structures, retreat to their regional strongholds, "considering it better to reign in hell than serve in heaven...The rallying cry of the majority is centralisation, while that of the minority is autonomy" (Michels, p198). The potential creation of small regional oligarchies does not, however, detract from the basic thesis that oligarchy is a function of organisation and is inevitable.

The validity of Michel's "iron law" has been investigated by numerous industrial relations commentators. Empirically based research (Lipsett, Leiserson, Hartmann) appears to confirm the general thesis of entrenched leadership. Other, more theoretically based research (Banks, Hyman, Coates), questions the validity of focusing narrowly upon this variable or on the alleged role of education associated with it.

Banks, for example, concedes the point that the departure from primitive democracy and amateurism in the name of effective, trade unionism carries with it unintended consequences. He, nevertheless, argues that it is wrong to assume that representative democracy (or the historically specific form of delegate democracy within the NUM) has no check on oligarchic tendencies. On the one hand many trade union officials act in a delegated, mandated capacity voting and acting only according to instructions of the rank and file. On the other, where leaders do have representative status, they are accountable to ordinary members for actions taken. The fact that they are, generally successful in elections is not, in itself, evidence of oligarchy. They may retain office because of their abilities and political skills, skills which become necessary only because democracy exists.

In making this point Banks focuses on the role of an intermediate strata of leaders or aspiring leaders "whose challenge must be met and whose political skills are not negligible" (Banks, p90). Hyman echoes this point and argues that Michel's analysis is limited to a consideration of the formal, national channels of decision making and ignores the importance, particularly in a British context, of shop-floor trade unionism. Active membership participation at this level is very much in evidence and "while workshop organisation normally has no formal relationship to the official structures of trade unionism, in practice the shop steward represents a crucial link between the membership at large and the union hierarchy" (Hyman).

Whereas Hyman views shop steward influence as an actual demonstration of democracy, others (e.g. Batstone et al.) suggest that the phenomena, in Britain, of full-time shop stewards utilising a variety of company provided resources such as offices and "check-off" facilities for the payment of union subscriptions, may result in stewards having a degree of dependence on management such that "in certain respects it is possible that there are processes occurring at shop-floor level similar to those Michels and others have identified at national level" (Batstone et al., p2).

This is confirmed by Hartmann who, in broadly confirming the Michels' thesis, conducts his analysis at the level of Works Councils in Germany, thereby emphasising the importance of analyses which focus upon different levels in the trade union hierarchy.

In their focus on leadership "immunity" then, the above arguments cast some doubt on the Michel's thesis in so far as the attention is restricted to an analysis of national level leadership. Nevertheless, in highlighting the role of alternative levels in the trade union hierarchy, they suggest that there is sufficient force in the thesis to conclude that what has occurred in the departure from primitive democracy is, at least, the creation of a "polyarchy", a balance of power between the leaders and a minority of active participants who occasionally challenge their authority. The larger, more active and more responsive this group to the rank and file membership, the closer the organisation approximates the democratic ideal. The smaller the union, as the NUM is becoming, then "the more the organisation moves towards the oligarchic pole" (van de Vall).

Coates' critique of the "iron law" differs from the above. For Coates, Michel's thesis is relevant only in the context of "a labour movement based upon fairly limited levels of popular literacy" (Coates, p81). Today, he argues, many more qualified candidates are available for office and the relations between "labour professionals" and those they serve has been transformed by increased education of the rank and file. Thus, "the Iron Law of Oligarchy" is perhaps a fair description of a bygone division of social labour in which the clerk was a natural aristocrat: but today's skills have certainly modified the landscape with which Michels was familiar" (Coates, ibid).

Banks, too, focuses on the alleged role of education in distancing leaders from led. He questions the relationship between formal education and leadership positions suggesting that the education which is important may not be the "intellectual" kind attained at schools and colleges/universities. Rather it is that borne out of experience on the shop floor or "coal-face" and which continues "through experience as a shop-steward, a collector of subscriptions, a branch secretary, administrator of local union affairs on an unpaid amateurish basis before eventually becoming a part-time and then a full-time officer" (Banks, p91).

He concedes that there is an increasing tendency for trade unions to employ college/university graduates, but suggests that they are likely to be deployed in subsidiary roles. They occupy research posts, edit journals, act as contact men with government officials, sit on committees of enquiry and generally take responsibility for public relations. Studies of this kind of "expert", according to Banks, shows that he "typically has consistent sustained influence on problems that the top officers see as far from the core function of the union... ...areas where direct experience of work is not seen as relevant" (Banks, opcit).

Importantly, however, this kind of individual is appointed not elected. S/he is responsible to the trade union leaders and the formal, "subsidiary roles" that they serve are directly to this leadership. S/he, it may be hypothesised, is likely to possess qualities, attitudes and ideologies consistent with those who are responsible for their appointment. The extent to which they become an integral part of any oligarchic tendency or act as a bulwark against such, perhaps strengthening an emergent polyarchy, may be a necessary part of a research agenda.

From the perspective of ordinary members then, representative democracy, whilst promising benefits of increased efficiency and bargaining power, carries potential costs of disenfranchisement as member's roles and those of elected officials acting on their behalf became so differentiated and separated by intermediaries that direct communication and participation in the decisions affecting them no longer seemed possible. In short, the rank and file are faced with the dual problem of how to control the union's executive and how to control the administration appointed to serve them. This issue is now taken a step further in the context of the widespread introduction of IT in organisations.

Orlikowski, in examining the introduction of IT in organisations, makes significant use of Heydebrand's concept of "technarchy". Technarchy is described as an administrative control structure encompassing technocratic forms of organisation and containing a number of important elements including, a fusion of managerial and professional/technical roles; extensive use of systems analysis, planning oriented towards crisis prevention and, most important in the current context, the primacy of technical rationality over democratic and social values.

In technarchic structures the "system" is pre-eminent. It incorporates and transcends authority based on bureaucratic and professional criteria such that "all are subordinated to its imperatives" (Orlikowski).

IT facilitates the implementation of technarchic strategies in a number of ways. Firstly, it is seen to "de-bureaucratise" partly in the way in which new office information systems cut across the boundaries of organisational units that were formerly autonomous and partly because IT can reduce horizontal and vertical links via the installation of "intelligent workstations" (Orlikowski op cit).

Secondly, it is seen to "de-professionalise" administration and control in that the autonomy and authority of professionals is eroded by those who access computers that store and handle knowledge that was previously the preserve of the professional. This is seen to be particularly relevant in the case of codifiable knowledge generally obtained by professionals via an academic education.

Thirdly, in reducing complexity, unpredictability and inconsistency and replacing it with standardisation, rationalisation and routinization, computers and computer processed information become "icons of rationality". They and those who operate them exude a technical rationality which is elevated to the status of appearing as "the only form of reason, the only arbiter of rationality" (Orlikowski, p26). In so doing, the "system" can constitute a new, dominant ideology, which although less obvious than previous ideologies is "more invidious and less vulnerable to detection, reflection and resistance" (Orlikowski, ibid).

These, rather negative consequences, are associated with particular systems and it is conceded that some systems are "less atomistic and make provision for group interaction and collaboration". It is also suggested that some aspects of the introduction of IT ought to be seen in a more positive light. End user computing, for example, is said to provide the users with a computer capability which allows them to develop and operate their own systems and intra-organisational computer networks allow for the spontaneous and rapid transfer of data and communications (via electronic mail and facsimile transmission).

Additionally, in the context of a labour process analysis and taking a pro labour perspective, the "de-bureaucratisation" aforementioned and the decentralisation implied in this, carries the danger (for managers) of information dispersal, functional disintegration, an inability to control users' development of systems and the inability to manage computer networks.

In addition, Orlikowski notes that Habermas questions the alleged dominance of technarchic systems suggesting that the capacity for such is limited since "cultural and communicative systems of meaning remain resistant to technocratic domination" and attempts at subordination are likely to "engender a crisis of legitimation through unintended effects of re-politicisation and the development of participation and alternative forms of life and work...".

It is clear from the foregoing, that much more could be said in the context of the "technarchy" thesis than is possible in this paper. It is equally clear that the prime relevance of its focus is in the context of capitalist/labour relations and, as such, it promises a novel and vital contribution to the extensive labour process literature. Nevertheless, it has a potential application in our chosen context of the introduction of IT in trade unions and raises issue which, we believe, have not yet become a part of this literature.

The nature of the issues and how they may be translated into a research agenda have yet to be determined. For the time being we simply make the following point. IT and the technarchic structures to which it allegedly gives birth are yet another factor to be taken into account in the context of analysing trade union governance. As the foregoing demonstrates, IT offers at least the potential for creating a new and different kind of authority and control structure within organisations. Yet it is not clear whether any such developments exacerbate or dilute existing tendencies towards oligarchy/polyarchy within trade unions. Whatever the outcome, it is important to guard against adopting a "technological determinist" stance. The way in which IT affects the decision making processes in trade unions will be determined by the choices made by individuals both within the hierarchy and within the rank and file, though clearly such choices will be constrained and influenced by the general economic and political environment faced by the union. As will be seen in the next section, this is particularly the case for the National Union of Mineworkers.

2 The NUM - the ecomomic and political context

The National Union of Mineworkers

According to Edelstein and Warner, "The British Mineworkers' union combines perhaps the highest degree of regional administrative autonomy in Britain... ...with national decision making on industrial and political policy.", thereby constituting "an important exception to the iron law of oligarchy" (Edelstein and Warner, p209).

The NUM established in its modern form in 1945 with over 600,000 members, is a federal union comprising 12 areas, nine of which are geographical with three based on occupational divisions. It is an industrial union that organises only in the coal industry and represents a majority of both surface and underground manual workers. The white collar workers, organised in COSA (Colliery Officials and Staffs Association), are affiliated to the NUM.

The union, ostensibly with a current membership of 116,000, organises at three levels: Branch, Area, and National, with a broadly similar structure at each level.

Branches are based on pits or other British Coal work places. Each Branch belongs to one of the NUM Areas. The Branch elects a committee and officers one of whom represents the Branch at NUM Area Council meetings. Two officers and half the committee come up for election every year. Participation in branch elections has traditionally been high. The branch plays an important bargaining role in such areas as manning and health and safety. Branch representatives are recognised by the employer regarding certain procedures. Independent industrial action at branch level is a regular occurrence.

The Areas have a great degree of autonomy with their own rules, finances, full-time elected officials, and staff. They are governed by Executive Committees with the larger areas having an Area Council consisting of delegates from each branch. Each area is represented on the National Executive Committee.

Currently in the NUM there are three national officials, the President, Vice President and the General Secretary. The President and General Secretary in constitutional terms, are of equal status. They are elected for five years in a ballot of the entire NUM membership. They run national office with a number of departments e.g. Industrial Relations, Research, Finance etc. staffed by full-time appointed officers. The President chairs meetings at Annual Conference and the National Executive Committee (NEC), carries out duties entrusted to him by the latter and ensures that the union is run according to its rule book. The General Secretary is the treasurer and conducts correspondence for the union and its National Executive and maintains records of their proceedings. Both officers are ex officio members of the National Executive. Neither has a vote on the NEC but the President does have a casting vote.

Annual Conference comprising delegates from each constituent Area is the supreme decision-making body of the union. Responsibility for the implementation of conference decisions rests with the National Executive Committee. This consists of representatives, usually full-time officials, from each Area elected every two years. The larger the area membership, the greater is the representation on the National Executive.

The coal industry: an historical overview

The coal industry was nationalised in 1947. During the following decade the industry was able to sell all that it could produce. However from 1957 with increasing competition from oil the industry began to contract significantly in terms of output, pits and manpower with little concerted resistance from the NUM, though with considerable and sporadic industrial action taken at branch level (Hall; Fine and Harris, Ch6). The major increase in the price of oil in the 1970's which looked set to boost the fortunes of the coal industry, merely slowed down the rate of contraction. Thus between 1947 and 1978/79 output fell from 188m tonnes to 106m tonnes with the total number of workers in the industry declining from 718000 to 297000 (British Coal, Annual Reports and Accounts). However in the 1970's miners did become more militant. There were two major industrial disputes concerning wages in the industry in the early 1970's.

In 1979 the Conservative Party came to power in the UK. The new, Thatcher government took office with a commitment to improve upon a relatively poor economic performance allegedly caused by over powerful trade unions and excessive State intervention. Given the aforementioned disputes of the early seventies and their perceived role in defeating the then Conservative (Heath) Government, the portents for the mineworkers were not good.

In essence, the Government's intention was to expose the coal industry to the rigours of the market, both domestic and international. Taken alongside Government policies to reduce trade union power through legislation and other means, the push by British Coal to increase productivity, and the stagnant market for coal, the result was to reduce significantly the strength and bargaining power of the NUM.

Resistance to Government policy did, nevertheless, occur, the most important example being the 1984/85 strike by the miners. This was provoked in March 1984 with the announcement by British Coal of major reductions in capacity and pit closures. The strike went on for a year and ended in a major defeat for the NUM.

The industrial action was fatally weakened by miners continuing to work particularly in the Nottingham Area. In the midst of the strike, a breakaway union, the Union of Democratic Mineworkers, was established with support from British Coal management. (MacGregor, pp219, 220, and 338). NEC members opposed to the strike argued that a national ballot should be held in accordance with the constitutional requirement for a 55% majority to declare a national strike (Winterton & Winterton, pp 70-72). As the then Vice-President put it, "We shall not be constitutionalised out of a defence of our jobs. Area by Area will decide and, in my opinion it will have a domino effect." (cited in ibid, p70). The NEC decision not to hold a national ballot was seen by opponents such as the Chairman of British Coal as a "high jacking of democracy from members of the union by the leadership" (MacGregor, p179). Even sympathisers thought the policy wrong, though for different reasons. Some saw it as a sign of weakness (Winterton and Winterton, p70). Others as a tactical mistake, (Beynon, p7).

The strike was described as "a landmark in the political and economic development of post-war Britain." (Beynon, p1). The Government welcomed and accepted the confrontation "...in the belief that success....would weaken the unions and activists, affirm managerial authority and advance the policies on which employers and governments are bent." (Miliband, p126).

Management strategy in the aftermath of the strike

The miners returned to work defeated, demoralised and divided. Dissension developed within the NUM both at Area and National level. The way was now clear for the Government and a more confident and aggressive British Coal to implement their cost cutting strategy. As Ian MacGregor the chairman of the National Coal Board put it at the end of the strike, " People are now discovering the price of insubordination and insurrection. And boy are we going to make it stick." (cited in Feickert, p10).

This strategy was to be achieved by: accelerating the rate of pit closures and concentrating production on low cost pits and open cast sites; abolishing old working practices and introducing labour flexibility e.g. extending the working day and working week; systematically reducing jobs in every part of the mine; increasing the use of sub-contractors; more extensive use of technology; and by decentralising pay bargaining from national to area and then to pit level (ibid, pp11-14). The coal industry is now being prepared for privatisation.

The effects of this strategy can be seen in table 2 which compares the six years leading up to the strike with the subsequent six year period. There was an acceleration in the rate of contraction in the number of miners and pits and in the membership of the NUM after the strike. From 1984 to 1989 the number of pits fell by more than half, the number of miners and NUM members by more than 60%. While output declined at a similar rate in both periods, productivity on the other hand soared by 78%.

Table 6.1
Change in pits, employment, output, productivity, NUM membership 1978-1989

	1978-1983		1984-1989	
No of pits	-53	-24%	-96	-57%
No of miners	-41900	-18%	-106000	-62%
Output (deep-mined)	-15.4mt	-15%	-14.5[c]	-16%
Productivity[a b]	24t	+8%	+1.89t[c]	+78%
NUM membership[b]	-13463	-4%	-194605	-61%

[a] overall output per man shift

[b] figures relate to 1983 to 1989 as figures for the strike year of 1984 are not available from the Certification Officer. Total membership includes a number of special categories (e.g. honorary, retired, unemployed) who are members but who either pay no subscription or make a nominal contribution only. Thus the change in NUM membership should only be taken as a very approximate indicator. In 1989 British Coal employed 85,000 employees while NUM membership in that year was 123,000. NUM sources estimate that the union currently has 42,000 members actually working in the industry.

Sources: British Coal, Annual Reports and Accounts; Annual Reports of the Certification Officer.

[c] 1983-1989

The NUM and the 1990s

At the time of the strike the NUM comprised more than 300,000 spread over 19 areas. Since the conflict membership has fallen to 123,000 in 12 areas. It is inevitable that the organisation of the NUM will undergo further considerable restructuring in the 1990's given the significant threats posed by the likely changes in the internal and external environments of the coal industry. Management will persist with its offensive. Productivity, as a result, will continue to increase rapidly. British Coal has estimated a worst case scenario where consumption of coal by the largest consumers, the electricity generators, could fall from the current level of 65m tonnes to 20m tonnes over five years. Gas will be increasingly substituted for coal in electricity generation: British Coal estimates that up to 30 million tonnes of coal sales could be displaced by gas (The Guardian, July 23 1992). Stricter environmental regulations within the European Community (e.g. the EC Directive on emissions from large combustion plants) will cause problems for the large traditional coal-fired power stations with a consequent threat to their demand for coal. N. M. Rothschild, the merchant bank advising the government on coal privatisation, has identified only 12 of the current 48 pits as profitable (Financial Times, May 19 1992).

Such threats to the future structure of the union are recognised by the leadership who are now taking a more proactive approach to restructuring. The post of General Secretary is to be combined with that of the President; merger talks have already taken place ,unsuccessfully, with the Transport and General Workers Union; there has also been a move, initiated at national level, towards more centralisation, the proposal being that head office take over the Yorkshire Area with the posts of National and Yorkshire Presidents being combined. While Area officials were in agreement the proposal did not find favour with Branches in Yorkshire as it would require them to surrender a significant degree of financial and decision-making autonomy. Thus structural changes are emerging within the union, and these may be seen to create further tensions which will test the extent to which the union can, in Edelsteins' view, remain an "important exception to the iron law of oligarchy".

3 Informatiom technology and the NUM

This section examines the role of information technology (IT) in the NUM. For the sake of clarity, however, it is important to emphasise a number of aspects concerning the peculiar focus adopted.

To begin with, although the focus here is somewhat different, we are, nevertheless, aware that a prime motivation for the adoption of an IT strategy in trade unions in general, as with other organisations, will be an attempt to attain organisational goals more "effectively". Whereas in most organisations such goals can reasonably objectively be determined, however, the nature of trade unions generally precludes such an assessment. Thus we neither know nor make any attempt at objectively defining what the NUM's goals are, though a pre-occupation with sheer survival would hardly be surprising given the political and economic constraints.

Secondly, whereas empirical research may be expected to establish subjectively defined goals (e.g.. by a survey of members) and to analyse the appropriateness of a particular strategy for the attainment of such, we have not conducted such an exercise.

Thirdly, there is no attempt made to analyse in any sophisticated form the precise nature of the IT strategy adopted by the NUM nor the extent to which this is consistent with the union's goals.

Important as all such foci are, they are not the concern of this paper. Rather, the focus in this section and the stimulus behind the paper, is the attempt to analyse the processes involved in introducing an IT strategy into the NUM and the role played by particular actors in this process. This, it is hoped, will shed some light on the implications for democratic decision making processes of introducing It into trade unions.

The information on which the section is based was obtained by "casual observation". It arose as a "by-product" of a training and consultancy contract during which the union's objectives and the need for an IT strategy to prosecute such were taken as "given". Nevertheless, and as one would expect in the context of consultancy, information regarding the purpose of introducing an IT strategy was sought by the consultant and the relevant costs/benefits and advantages/disadvantages of its adoption were reported to the appropriate decision makers in the union. Only in the aftermath of the consultancy and in consideration of the peculiar roles adopted by some of the actors involved, was information sought on the nature of the decision making processes, the identity of the "initiators" of the strategy and the communication channels through which information is both processed and "filtered". It was the answers to such questions which prompted the peculiar focus of the paper and which determines the nature of this section.

The section details and analyses various milestones of decision making by the NUM on the acquisition, use and implementation of Information Systems based upon IT using desktop computers. By its nature, it is descriptive, merely recording the chronology of the decisions leading to the introduction of IT and highlighting the role of particular individuals and groups in this process. The focus is on the period from the late 1980's until late 1991.

Until the recent investigation into the merits of an IT strategy, the extent to which the National Office of the NUM had embraced IT was limited to the purchase, during the 1980's, of an ICL computer to support finance (one of the most common justifications for the acquisition of IT), several Rank Xerox word processors for secretarial support, a personal computer in Social Insurance, some electronic typewriters and a computerised telephone system.

In 1987 the Chief Executive asked Thames Information Service (TIS) to provide an IT Strategy for the National Office. TIS, who advised on IT strategies for a number of unions, issued a report recommending capital expenditure of £280,000 and annual costs of some £50,000 for the implementation of a multi-user system rather than a network of microcomputers. The proposal was rejected by the leadership ostensibly on cost grounds.

Increasing pressures on declining numbers of National Office staff, (consequent upon a policy of maintaining a strict ratio of staff to, declining, membership) and the requirement to operate with obsolete technology led to both a loss of morale and a certain loss of administrative efficiency. To the extent that this occurred, however, it appeared not to concern the leadership and there is no evidence of it having been discussed at formal or even informal levels. It certainly never appeared as an agenda item on either National or Area Executive Councils.

Instead, the need for an IT strategy, became an issue first raised by Research staff at the National Office, who argued in favour of Decision Support Systems for the "political" activities undertaken by the Industrial Relations, International and Research Departments. The request was for a networked system which would facilitate both intra-departmental and inter-departmental communication, providing access to information of importance to all departments. There were also attempts made by Branches to obtain a networked system, with two Yorkshire branches, Grimethorpe and Armthorpe, moving resolutions at an Area level for a national IT initiative. All such initiatives went unheeded by National Office.

Leadership intransigence on the issue of IT in general was both well known and long standing. Indeed, in the 1970's the, then, National Coal Board (NCB) had approached the National Office of the NUM with a view to granting networking facilities to the union for £10,000. Despite apparently offering, inter alia, the prospect of providing the NUM access to NCB information vital to the conduct of collective bargaining, (a gesture hardly likely to be repeated in the current political environment), the leadership rejected the offer. Whether this was simply latter day "Luddism", a general alienation-ignorance of the potential benefits to be gained from embracing such facilities or merely a mistrust of "a hidden agenda" on the part of the employer is not clear.

Indeed, the unauthorised purchase of two computers in the late 1980's by an appointed national official led to the sequence of events which led to the setting up of the Lightman inquiry by the National Executive in 1990. The purchase resulted in the official resigning, followed by a complaint to the police by the President of the NUM (Lightman, p109).

Persistence on the part of the Research department eventually brought some positive reaction from the leadership in 1989 when the Work and Employment Study Group (WESG) at Leeds Polytechnic was commissioned to investigate the computing requirements of National Office staff. The appointment of the WESG was based on a knowledge of their preference for a participative approach to systems design.

This participative approach involved all the employees at National Office not only providing staff with a primary role in determining their operational requirements but also ensuring that a draft report and recommendations would be issued for their approval and the final Report not submitted until they were wholly satisfied with it as a specification of their requirements.

With the exception of the Finance department who had only recently become familiar with their enhanced ICL based system and were somewhat reluctant to undergo further change, all other sections were positive both towards the investigation and the participative approach to the determination of their requirements.

The final Report was sent to the National office indicating that a Local Area Network supporting word processing, a database management system and a spreadsheet package would satisfy their needs. Furthermore, WESG emphasised the need to go to tender to meet both the software and hardware requirements. The report was agreed by the staff of all sections involved.

None of this, however, amounted to an IT "strategy". Rather, it constituted an approach based on a series of ad hoc purchases of "bits and pieces" which, although, undoubtedly, efficiency enhancing, could never hope to achieve anything approaching the potential benefits to be gained from a properly networked and integrated system. Yet National and Area level officials appeared resolute in their determination to prevent the introduction of a National strategy partly, it would appear, because of a fear on the part of Area officials that a National IT strategy would lead to an integration of financial records thus undermining their source of authority in the Areas. Rees and owen found that local activists in the TGWU saw the computerisation of membership records leading to a ,loss of local autonomy (p28).

As a consequence of this, the National office staff in the various departments in Sheffield requested a meeting with the WESG IT analyst and commissioned a report on the merits of an IT strategy. This was done and the Report, which itemises in some detail the adverse consequences for the union of not evolving an IT strategy given the particularly threatening environment; the benefits measured in terms of increased efficiency, an improved quality of working life and the synergy of the resulting infrastructure, is reproduced as Appendix A. The National Executive took no action.

By the latter part of 1989 the secretarial staff at National office were somewhat frustrated at the increasingly "obsolete" nature of the word-processing equipment. The Research Staff increased the pressure on the National leadership by refusing to carry out instructions to investigate machine replacement unless and until a commitment was given to investigate seriously a proper integrated network system. Although the staff received such a commitment, the final order excluded, for reasons of cost, the crucial components to enable networking. That is to say, the order only contained several independent microcomputers.

This is not to suggest that all officials were as intransigent as the National leadership appeared to be. Indeed, some Area officials, for example, in the North East Area, had commissioned reports and invited tenders for the implementation of a network system at exactly the same time as the National office appeared content to spend considerable sums of money on non-networked equipment.

The Research Staff again commissioned WESG to submit a Report both on the merits of a fully integrated network system based on the national membership database and a critique of the, now installed but poorly designed, membership database system adopted by the North East Area. The intention was to use this as the basis of a Report to be presented by the Research Staff to the National Executive Committee (NEC) of the union as a justification for developing an IT strategy.

Despite the submission of this Report (see Appendix B) the IT strategy never materialised but by Nov/Dec 1990 the National office at Sheffield was networked and the Research Staff had also extracted firm commitments from National office and NUM North East for IT training (to be carried out by WESG using their preferred Action Learning approach to IT training). This was followed by an indication from NUM Yorkshire that they would consider a similar implementation.

A member of WESG was given a 3 months secondment with an overall remit to design and implement an appropriate training programme for the staff at Sheffield (National Office) and Durham (the North East Area H.Q.) which involved:

1 carrying out a Training Needs Analysis in word-processing for each office,

2 implementing the word-processing Training Programme at each office,

3 undertaking a network management analysis at the Sheffield and Durham Offices,

4 implementing a risk management analysis at the Sheffield and Durham Offices, and

5 implementing a detailed Information System Analysis of the compensation systems at NUM COSA, Doncaster and NUM NE, Durham.

Consistent with the WESG philosophy of "participative systems design", one element in the training package was an attempt to encourage the users at each of the offices to set up their own liaison with the equipment vendors, the intention being that they then become responsible for their own "system" and, in the process, increasing their self-esteem, independence, learning and control over suppliers. However, the staff at all offices were, in their view, under too much pressure to take on such responsibilities or to accommodate, what they thought would be the steep learning curve required to act as an effective liaison.

There were two NUM officials requiring network management training.

One, the President of the North East who had been the enthusiastic initiator and supporter of the network in that Area in 1989. The other was the Assistant National Research Officer at Sheffield. Given the rapidly changing and rapidly deteriorating political and economic environment aforementioned, neither was able to commit as much time to the Training Programme as formally envisaged. As a result most of the network management was carried out by the WESG members who established an interactive training course on the two prospective network managers' work-stations to enable them to follow up the training whenever possible.

Consequently requests were made for Training Programmes, similar to the above, to be devised for Branch officials. Negotiations regarding the form, timing and costing of the Training Programme were undertaken between 2 elected Area officials, 2 appointed Area officials (the Heads of both the Finance and Industrial Relations sections) and the WESG member. Whilst agreement was reached on form and timing, costing proved something of a stumbling block, primarily because of the Area Executive Council insistence on, though refusal to countenance the expenditures involved in, a minimum of 4 lay officials per branch attending training sessions during normal shift hours.

Suggestions from appointed National office officials and the NUM Lecturer for either half the number of officials and/or training during non-work hours has been met with a blank refusal. In the opinion of the Assistant National Research Officer, the NUM Yorkshire feared that Branches could use such technologies to take independent and unofficial action, which in the context of current industrial relations legislation could be litigious.

In June the north eastern Regional Manager of British Coal (BC, used to be NCB) sent a letter to the General Secretary of the NUM NE, Durham stating that BC would not be running check-off for the miners at Monkwearmouth pit as there had been a local dispute between the miner's Lodge (NUM NE's nomenclature for a Branch) and the pit manager for some time. (Check-off is a system whereby the employer agrees, usually for a fee, to deduct the union subscriptions from the union members pay on the union's behalf sending the total to the union.)

The WESG member devised clerical procedures to support the move towards both cash collection and Direct Debit. He later specified the requirements of NUM NE to a computer consulting firm which had agreed to enhance their BAS-E to meet them.

The Treasurer and President of NUM NE pressed for the men to pay their subscriptions by Direct Debit to make the collection more economic. The General Secretary, on the other hand, argued for cash collection at the Lodge as it would increase the contact of the lay officials of the Lodge with its membership thus politicising the Lodge activities. To date 1:4 members pay by Direct Debit.

4 "Technarchy" in trade unions prospects for a research strategy?

The purpose of this final section is to pose, albeit rhetorically, the question of whether, in the context of an information technology age, we might be witnessing the emergence of new administrative control structures within trade unions and to suggest a research strategy for analysing the implications of this for the democratic decision making processes therein.

There are basically two interrelated and logically sequential aspects to a research project analysing the existence, form and influence of "technarchy" in trade unions. The first, is clearly to identify the extent to which technarchic processes exist. This involves an analysis of the framework "technarchy" suggested by Orlikowski and the particular form which it may take in trade unions. In other words, we need to examine the relevance of and extent to which, in trade unions, there has been a merging of professional/technical roles; a widespread adoption of systems analysis and "systems thinking" giving information technologies a "pivotal and constitutive role" in decision making ; an increased prevalence of crisis management techniques, anad, most importantly, evidence of technical rationality taking precedence over democratic and social values.

Secondly, it is necessary to identify the key individuals/ groups who formulate IT strategy within trade unions and examine their power and influence.

There are, undoubtedly, considerable practical difficulties in attempting the above. There are also considerable conceptual problems not the least of which concerns the nature of "power", its sources and how we might identify its use. This is briefly discussed before suggesting a research strategy.

The essence of the oligarchy thesis appears to be that the "formal" distribution of power need have little relationship to the "real" sources of power in voluntary organisations. In other words the very identification of the source of power is itself a major problem. Despite this, Michels, and others testing the thesis, focused on formal levels in the trade union hierarchy, examining indices of "immunity" and/or the closeness of elections for formal positions in the hierarchy. As Batstone et al, op cit argue, it is unlikely that such foci can tell us much about the use of power. Rather there is a need to analyse overall patterns of behaviour and the contexts within which they occur, that is to say at actions rather than formal responsibilities.

Thus a definition of "power" should not be related to formal positions of authority as embodied in rule-books and/or constitutions. We need to adopt" a more sophisticated concept of power and attempt to employ methods which permit recognition of its various dimensions" (Batstone et al, op cit, p270),

At its most simple, "power" refers to the ability to achieve one's goals and to persuade others of the merits of particular arguments or to impose your will in the event of a failure to convince. Power, thus defined, may be most visible in the context of decisions made since these are either agreed upon or unilaterally imposed. Thus, as with the NUM case study, we could look at the actual decisions as constituting outcomes of the actions or patterns of behaviour of individuals/groups. By itself, however, this is a very partial indicator. A focus only on decisions made or even on issues discussed, ignores power which is exercised to prevent other issues reaching the agenda. So although it is important to see who the initiators of issues and actions are, we need to look beyond these to see who, how and why certain individuals and groups have the ability to influence the "terms of the debate".

A further source of power is the ability to convince others of your competence to decide without discussion. This may be done on the basis of particular skills and competencies of a technical, academic or personal nature. Of course, this is a very complex aspect of power since, by its nature, it makes it difficult to ascertain whether and to what extent "significant others" influence the decision. This makes it important to understand individual's frames of reference and ideological perspective.

A final potential source of power in the organisation may be the extent to which the latter itself occupies a central position in determining members terms and conditions of employment. As Batstone et al suggest in the context of shop stewards, where the union plays a key role in determining the experience of members "such centrality leads to the "mobilisation of bias"... "a set of predominant values, beliefs, rituals and institutional procedures that operate systematically and consistently to the benefit of certain persons and groups at the expense of others" (Batstone et al op cit).

In analysing power and influence then, a knowledge of ideology and institutions are as important as a focus on decisions made and/or issues discussed, or those which aren't and we need to identify:

1 Who actually makes the decisions.

2 Who identifies and shapes issues for decision and influences the subsequent decisionmaking processes.

3 Who plays the major roles in creating and/or maintaining the larger ideology and perspective.

In researching such aspects it is important to recognise the vital role played by social networks, how these operate in specific unions, particularly at the level of the "bureaucracy" and/or administration, and how these both connect and are connected to the decision making process and the patterns of power and influence in the organisation.

A research strategy

The theoretical approach outlined suggests a focus on particular areas of study rather than the provision of specific hypotheses to test. It requires a research method permitting a high degree of flexibility and the maximum collection of data in a form enabling the researcher to understand the dynamics of decision making in the peculiar context of the organisation under investigation.

The above can only be undertaken by means of a case study research method based upon observation and the extensive use of interviews, questionnaires and interrogation of documentary evidence.

Batstone et al highlight the difficulties in such an approach particularly given resource constraints. Observation makes heavy demands on time; it needs to be carried out over a considerable period of time and often involves a considerable period of learning about the situation being observed. The range of situations studied need to be limited. However, comparative research may help to avoid a situation leading to the generation of ideas and concepts which are too "parochial".

The research process starts, quite properly, from an articulation of ideas and concepts borne out of experience or the reverse, an attempt to apply such to concrete situations. This paper takes the former approach. A useful research agenda could be compiled to undertake the latter.

References

Banks, J. A. (1974), *Trade Unionism*, Collier MacMillan, London.

Batstone, E., Boraston, I. and Frenkel, S. (1979), *Shop Stewards in Action*, Basil Blackwell, Oxford.

Beynon H.(1985), *Digging Deeper*.

Beynon H., et al,'Opencast Coalmining and the Politics of Coal Production', *Capital&Class* , Number 40, Spring 1990.

Braverman H.(1974), 'Labour and Monopoly Capital: The Degradation of Work in the Twentieth Century', *Monthly Review Press*, London.

Child J., 'Organization Structure, Environment and Performance: The Role of Strategic Choice', *Sociology*, Volume 6, 1972.

Clapham M., 'The local and area bargaining unit in the UK coal mining industry', in *The Industrial Tutor*, Volume 5, Number 4 Autumn 1991.

Coates K. and Topham T.(1988), *Trade Unions in Britain*, 3rd edition, Fontana Press, London.

Edelstein J. D., and Warner (1975), M.,*Comparative Union Democracy*, London, George Allen & Unwin.

Feickert D., 'The Midwife of Mining', *Capital & Class*, Number 31, Spring 1987.

Fine B. and Harris, L.(1985), *The Peculiarities of The British Economy*, Lawrence and Wishart, London.

Hall A.(1981), *King Coal*, Pelican.

Hartmann J., 'Works Councils and the Iron Law of Oligarchy', *British Journal of Industrial Relations*, 1979.

Hastings S., 'Report of the TURU Seminar on Computerised membership records' Oxford Trade Union Research Unit, Ruskin College, 1983

Hyman R.(1975), *Industrial Relations: a Marxist Introduction*, Macmillan, London.

Leiserson W. M. (1959), *American Trade Union Democracy*, Colombia University Press, New York.

Lightman QC, G.(1990), *The Lightman Report on the NUM*, Penguin.

Lipset, Seymour M., et al(1956), *Union Democracy*, Free Press, Glencoe, Illinois.

MacGregor I.(1986), *The Enemies Within*, Fontana.

Michels R.(1962),*Political Parties*, Glencoe.

Miliband R.(1989), *Divided Societies*, Oxford University Press, Oxford.

Orlikowski W.(1988), 'Computer Technology in Organizations: Some Critical Notes', in Knights, D. and Willmott, H. *New Technology and the Labour Process*, Manchester University Press.

Rees T and Owen D., 'The Computerisation of Trade union Memebership Records: A Research Note', in *Work Employment & Society*, vol 7, No. 1, March 1993

Salamon M., *Industrial Relations - Theory and Practice*, 1987, van de Vall, Labor Organizations, 1970.

Webb S. and B., *Industrial Democracy*, Longmans, London, 1898.

Winterton J. and R.(1989), *Coal, Crisis and Conflict*, Manchester University Press,.

Woodward J.(1980), *Industrial Organizations, Theory and Practice*, 2nd edition, Oxford University Press, Oxford.

Transformation of Patterns II
TECHNOLOGICAL DEVELOPMENT AND INNOVATION

7 Social environment of technical progress

Josef Hochgerner and Gerhard Berka

1 An approach to the problem

A world wide transformation is emerging comparable to the industrial revolution in the 18th and 19th centuries: A third industrial revolution has already been talked about. The so-called "micro-electronic" revolution is not only leading to discussions on new societal concepts in social scientific circles: On the one hand, the "post industrial society", either represented as a transition to a service oriented society connected to the breaking free from the classic industrial sector, or on the other hand the concept of the information society (information being a new productive force).

Various empirical data favour these concepts. In spite of an increasing contribution of information to the Gross National Product increase in the production of information, a change in the emphasis of the central production point in society is becoming clear: Already in 1982, it was estimated that around 50% of all employees in the USA alone were involved in either the generation, processing, or transmission of information (Hummel, 1982). In as far as Austria is concerned, a study exists for 1971 (Schmoranz, 1980) where 29.5% of all employed people were involved in information activities. A later study established that around 50% of the Gross Domestic Product as an average in the more advanced OECD (Organisation for Economic Co-operation and Development) member states were directly dependent on information and communication technology (Grewlich, 1986).

These data describe the situation since the seventies and make it so clear that the Dutch sociologist Hamelink stated: "The most important consequence which can be drawn from the current developments consists of the fact that a new telecommunication network is emerging which the future of Europe will, in one way or another, influence. What is disturbing however, is the complete lack of basic discussion over the direction and quality of this future. All groups which are directly involved are too deeply involved with their technology-driven short-sighted interests...Thus, Europe is blind, its path ahead is being steered by technical possibilities and wishful thinking" (Hamelink, 1986).

Other technologies as well are ascribed powers to transform society: The headline "complete automobile society" was introduced into the discussion in 1972 as a warning perspective (Dollinger, 1972). The transport system, based on the motor car, shows an astonishing growth. In 1988, individuals travelled in total 677 thousand million kilometres in Germany, 82% of which was by car (Seifried, 1990, 17). A similar picture can be seen in Austria. Up to the year 2011, a further increase of 50% is estimated in individual mobility whilst at the same time at least some observers expect a stagnation phase concerning public transport (ÖAMTC, 1990).

In the economic structure of the western industrialized states, the motor car no longer has an intentional significance. The density of vehicles in the world can quite clearly be seen (in 1986): 1000 Americans drive 556 cars, 1000 west Europeans have 352 cars whilst 1000 Africans own 9 cars and 1000 Chinese drive only 0.7 cars (Seifried, 1990, 42). The car manufacturing industry in the Federal Republic of Germany produced 4.6 million cars in 1987 though only 12% of the people employed in industry were involved (Seifried, 1990, 92). If one takes the trade and industry aspect into consideration, this figure becomes almost 20%. The share (measured according to activity) of the motor car industry in the individual branches (steel production, glass production, etc.) amounts to a figure between 15% and 30% (Armanski, 1986, 12).

Growing ecological threat potentials and the corresponding increasing sensitivity for the consequences of technology are obviously leading to a newly awoken interest in future societal models and considerations about the way in which every day life would change. Alongside the "information society" (information and communication technologies) and the "motor car society" (transport technologies), new key terms such as the "risk society" (nuclear energy) or "television society" (electronic media) have become famous. There is, however, a need for explanation for the sociology of technology: Being specialized in social scientific research into technology, answers are demanded which register societal perspectives above appraisals of the consequences of technology. The following question complexes remain in the limelight:

- The emergence of a societal form oriented towards technology (whatever the technology may be) presupposes the exploration of essential societal areas by the respective technology. How can this process be described both theoretically and empirically?

- The transition from an existing societal formation to a newly formed one also requires a transformation of the societal structural principles. How can such a transformation be described? Who could bear this change - by using what kind of societal concepts and instruments?

2 Evidence of mechanisation and societal transformation.

The somersaulting innovation cycles in computer technology, which are no longer measured in months, give the observer a feeling of rapid and incessant technological progress. Spectacular headlines in appropriate newspapers, however, cover up the fundamental and long term sociological trends which are "longue durée" as Braudel (1979) terms them.

In the history of communication, communication carried out by technological systems has created the exceptions which are limited to the last one hundred years. Until the nineteenth century, book and newspaper production for minority reception remained withheld, the quickest method of communicating was by using optic telegraphs, the technical principles behind which having remained unchanged from Roman times until Napoleon III. Only since the successes of Marconi and Edison in the second half of the nineteenth century has it come to incessant, even till today, thrusts of mechanisation in the communication of information.

The diffusion speed of communication technologies has increased enormously from innovation to innovation. Eighty years were required in order to establish one telephone into 37 million American households. For the same number of homes, the car industry needed only 49 years, and for television, no more than 10 years (Grewlich, 1986). Considering the development of radio and television, people living in the western industrialised societal formations find themselves amongst a network of technical means of transmitting information, through the use of which a large percentage of our daily information is transported.

Thus it is hardly astonishing that Grewlich (1986) has already spoken of a "global information economy". Information and communication technologies (like computer and satellite technologies) are linked to "telematics" through which a qualitative leap in development comes about: There exists a moment of simultaneity and ubiquity in which every place on the Earth can be reached with electronically transmitted information (which in theory is possible for everybody, though practically only valid for a socio-technical structured minority of the world population).

The motor car, even, as an integral transport technology can only glance back at a very short history: One can only really talk about a trend to motorisation since the end of World War I. By "motorisation" not only should the sale and distribution of automobiles be understood, but in fact all those preconditions must be taken into consideration with the car which elevate the idea of a car to a "motor car": "An automobile is, in itself, not enough; it needs many more roads in order to be driven upon...whilst other machines such as the type-writer or the vacuum cleaner lay no claims outside the four walls of the owners' house, the motor car demands that roads offer free driving and that people who do not have a car behave in a justified way to cars" (Sachs, 1990, 23).

The motor car requires roads which are built suited to car traffic, an infrastructure of maintenance (filling stations, garages and workshops etc.), and last, but not least, people to adapt themselves to the car. People were systematically ousted out of the street area of life (by pavements, zebra crossings, pedestrian underpasses, etc.) and forced to adapt their behaviour: "...it must be passed on to the habits of the population that where there are footpaths, one should not step over the kerb, when stepping from the kerb one should look around, one should walk and drive on the right-hand side of the road; even when the entire road is empty, one should not walk on the road" (Sachs, 1990, 44).

The extensive distribution of the automobile is beginning to take effect in its everyday usage: In urban areas, the average driving speed amounts to little more than between 20 and 30 km/h (compared to the bicycle 13 km/h). On motorways, the motor car reaches average speeds of 100 km/h (traffic jams being incorporated into the figures). If one takes the periods of time into consideration which must be included for the maintenance and running costs, and the external costs (environmental costs, etc.), a "social average speed" of 13-18 km/h results (Seifried, 1990, 68).

Due to the harm caused by road traffic, many individual cities are already beginning to distance the motor car from the urban conurbation. In Italy, Bologna has blocked its city centre from private road traffic. In Norway, a toll was imposed for car drivers wishing to drive into the Oslo and Bergen city centres. The city council in Zurich formulated the following principles as the foundations for new traffic policy: "Transport in the future is to be carried out using the most environmentally friendly means, rather than those which are apparently cheapest for the individual. Freedom of choice for means of transport must recognise its limits where the principles of life ... immediately or even in the long-term are seriously endangered".

All of which is evidence for a transition from a societal structure oriented towards the motor car to one freed from the motor car. Nevertheless this implies a radical change in the cultural system: The symbolic object of the motor car is losing its symbolic content and must accept the possibility of living on as an object of "negative" value in a new cultural environment.

The examples of mechanisation also demonstrate the possible "transitions", the carriers of societal transformations. In this instance of information technology the so-called "compulsory fact", i.e. the rationalization advantages generated by information technologies force industries beneath a competitive situation, and thus have to incorporate these technologies into the production process. The decision bearers function as carriers in production which leads into the radical change in society by controlling investment. The advance in information technology from production into the everyday world (computer, ISDN, etc.) is quite simply an economic consequence of changes in production. The consequences of this development (protection of data, "the invisible man", etc.) are presently more suspected than dealt with.

In the instance of transport technology, the initiation for the transformation comes from technology itself. With the growth of road traffic and the continued distribution of the motor car, the individual transport system is getting close to its limits of efficiency: With the daily traffic hold ups, the motor car is heading towards being a suitable means of transport ad absurdum. Even the current propagated attempts at repairing this situation by means of diversions and "park and ride" schemes up to the highly technical traffic directing systems all point towards a more even distribution of the hold up potential. The failure of technology is bringing about a social transformation from within.

Parallel to these examples, a potential of the societal transformation is emerging, unnoticed alongside the current ecological discussion, which hitherto has hardly been perceived: The technicians. An "ecological" picture could be placed opposite their until now traditional image ranging from a stone mason to a "mindless apostle of progress". It is a two-point question: Is a way really being paved in the ecological consciousness of all those active in technical professions for a "turning point"? Could this, assuming that technicians play a key role ("gatekeeper function"), lead to a change in direction of really "another" technology in the ecological and social sense?

So-called attitudes which are "critical of technology" are nearly as wide spread today in Austria's technical professions as they are in the entire population of Austria (Hochgerner, 1992): Three quarters of the interviewed technicians (a sample of 1042 taken throughout Austria) were able to handle the heretical opinion that "technology itself is a danger if people do not learn how to deal with it". For the claim that "wasteful packaging should be banned for all mass-produced goods", 77% agreed, whilst 69% wished to take their own future more seriously in their hands: "Technicians should appoint themselves more for controlling the consequences and for the humane form of technology". This incorporates the fact that a majority (around 60%) explicitly are in favour of "rigorous energy saving" and "cutting down traffic". Three quarters of the professional technicians who are themselves affected by these measures support the causing principle, and generally a further strategy aimed at improving control and openness.

All interviewees had either been working for five or for ten years. The striking transformation had taken place in the years during which they had been employed. This, and similar data, reveal a potential for technological and environmental policy discussion which, until now, has been generally underestimated and not been correctly understood. We will refer to this question on a more elaborated basis at the end of this article. But what does it take to provide a "more elaborated basis"? How are we going to explain sociologically what happens in this intertwined field of socio-technical systems? Are there certain patterns of change in society and technology we could analyse to produce comprehensive conceptions - practical enough to manage a socially and ecologically appropriate pace of technical progress? Does it take only "good" and ambitious (i. e. ecologically aware) technicians to make "good" (appropriate) technology? Or is there a need for a more complex understanding of what kind of social environment propels what kind of technological development (in particular ecologically and socially sound or rather dangerous and in the long run hazardous technologies)?

Europe has been the cradle of industrialization and thus of modern technologies, even though nowadays the most relevant impacts stem from the US and Japan, thus creating the triad of economically and technologically competing powers. According to the newly appearing demands on ecologically sound and socially relevant technologies, Europe now needs - as much as any other region in the world - to employ ideas and measures for a change of modes concerning the prevailing trends in the mechanization of production, transportation and communication as well.

3 The background to sociological conceptions of technology and the development of theories

Specific sociological research referring to technology, publications, as well as assessments concerning the theoretical and also the practical relevance of these activities all display a concurring component: The concrete social-scientific interaction and production of knowledge, even in the technologically most developed western countries, is not keeping up with the terrific progress in technology and the growth of social problems caused as a direct consequence of this. In fact "technology" has recently made a certain career from being the subject of congresses, books, articles and research projects. In different countries there are new institutions or research programmes specially developed for study referring to technology (particularly concerning the genesis of technology, the implementation of technology in work or profession, or further consequences of technology from a social or an ecological standpoint). Nonetheless a most extensive consensus that everything remains inadequately partly finished dominates this matter which is not only led back to quantitative financial restrictions, but essentially to a, as yet, undeveloped theoretical and methodical tool for sociological research.

Closer consideration of developments in social-scientific research into technology shows some new and promising trends. Principally philosophical discourse on the issue of the societal meaning of technology (e.g. Gehlen, 1957; Löwith, 1964), and authors like Marcuse (1941, 1967), Habermas (1968), and Linde (1972, 1982) lead to a lively debate in the German speaking cultures on the societal conditions of technological developments (e.g. Ullrich, 1979; Jokisch, 1982; Bammé et al., 1983; Rammert, 1983, 1991, 1993; Hülsmann, 1985; Beck, 1986; Hochgerner, 1986; Getzinger/Papousek, 1987; Hochgerner/Bammé, 1989; Tschiedel, 1989, 1990; Alemann et al., 1992).

On the other hand, a stronger argument supported by empirical research exists which emphatically deals with the raising of issues in the labour market and the sociological implications on professions - concerning particular professional groups such as engineers. Mostly these studies were as a matter of fact not conceived to be "contributions to the sociology of technology." However their results concern essential protagonists involved in determining the relative function and the shaping of technology, so that they have at their fingertips a wealth of information on investigations into these more recent special issues. Studies of this genre can be analysed in this new light just like Hortleder, 1973; Kammerer/ Lutz/Nuber, 1973; Langer 1981; or Janshen/Rudolph, 1987. The same is true for the classical investigation into the "consequences of technology" by Lederer (1938), and it goes without saying for the large number of industrial sociological studies of which only a few are cited here as being prominent examples: The more dated and more recent studies by Kern/ Schumann, 1970, 1984; Mickler et al., 1976; Altvater et al., 1985; Baethge/Oberbeck, 1986; Aichholzer et al., 1991.

These classically determined schools of thought, or even one of the already more or less long established main flows of relevant sociological research can also be interpreted in addition to their respective central meaning as being "technically sociological". Parallel to this, in the past ten to fifteen years at least four recognised new approaches have been established which bring about new important impulses which in turn develops itself successively into a common understanding of technical sociological theory and research:

- Standing in the limelight in many countries is the consolidation of the sphere of research into the social consequences of developments in technology and of the acceptance of research: Studies relating to the *"evaluation of technology"*, "technology assessment" and lastly "technology monitoring" (e.g. Nowotny, 1979; Radkau, 1983; Halfmann, 1984; Coenen, 1985; Jochem, 1988).

- Increasingly in the last few years, great importance has been put on research relating to the consequences of technology and, in particular, research into the *origins (genesis) of technology*. This corresponds to the widely accepted point of view that at best an individual technology should be analysed in the real meaning of sociological factors from its beginnings, which would determine the meaning of certain technological concepts within historically concrete societal formations. Important publications of this type include (to name but a few) Asdonk/Bredeweg/Kowol, 1990; Knie, 1991; Rammert, 1991; Dierkes/Hoffmann 1992.

- A third important trend can be outlined as a *perspective extension* of scientific awareness of socially relevant areas, which in the past had not been connected to the implementation and the effects of technology (in everyday life, private life, biology and ecology, agriculture, old age and health etc.). Several studies in this area stem from e.g. Zapf, 1986; Maimann, 1987; Braun, 1988; Joerges, 1988; Weingart, 1989; Wagner, 1991).

- Finally, in the Anglo-American countries as well as in Scandinavia the *constructive approach* is more commonly accepted: In technical sociological studies of this kind, the combined effects of the protagonists, interests, technological and scientific traditions are investigated in the process of finding a decision between different >technically< (in the original sense of the word, i.e. based on points of achievement and function) competing concepts of technology. Leading the way into this field is a book by Bijker/Hughes/Pinch (1987), followed for example by Cronberg (1991).

These different central issues in the contemporary sociological research into technology are naturally neither characterized by separation nor by basic exclusion. However in no way do they converge to form a common sociological theory of technology. More urgently though, they through more light on the question whether, and in what form or under what circumstances can such a theory exist. This often enters discussions or written articles in the shape of limitation and reservation of expression: Whenever one is talking about the "sociology of technology", one often adds that such a concept just happens to be in a birth phase. Attempts at integration are provisionally made with some degree of reservation.

The state of sociological research relating to technology paints a fragmentary picture. Several empirical studies carried out on grounds of our own theoretical conception are not excluded from this (Berka, 1992a; Hochgerner, 1992; Ornetzeder, 1992) even when we have always tried to come closer to a common concept in our theoretical publications (Hochgerner, 1986, 1989, 1990a, 1990b; Berka, 1992b). All in all, the numerous attempts and experiences alongside the available material which is frequently merely implicit, almost inevitably arouse an impression of unclarity.

Thus in this situation, how can it be possible to compose a compact theory which is calculated on a minimum demand on "robustness" (Rip, 1987)? If the future birth of new scientific knowledge is thought of as being dependent on the accepted body of scientific methods recognised in the relevant "scientific community", then no generally acknowledged theory relating to the sociology of technology can be expected from a lack of agreement and hardly any available consciousness of a corresponding "community". For the time being, the accuracy of an extensively secluded existing theory is excluded, considered as small now as its later efficiency. However both demand an extension of scientific opinion, hypotheses and the resulting acts of social interaction; to a certain degree a final acceptance of connecting theories relating to the sociology of technology will emerge as a result of social processes.

4 The need for theoretical foundation of sociological research into technology

The fragmentation problem does not exclusively concern flows of theories relating to the sociology of technology, but can be regarded as a symptom for sociology taken as a whole. Signs are emerging which are pointing towards a point of complete stagnation in the development of theories. The initial theories resulting from criticism of structural functionalism as they were formulated by Goffman (1970), Garfinkel (1967) et al., took shape from original ideas. Though they were engaged in a fight against the "Grand Theory", their radical beginnings forced, however, exactly a new claim for a monopoly theory. The confrontation with other competing theoretical programmes which emerged as a direct result of that claim brought about the question of the comparability of theories.

The 1970s were marked by the attempt to find a new "paradigm" - similar to the "Grand Theory" - and to uncover new possibilities to compare theories. Success was limited (e.g. Opp, 1970). Finally, the situation was rescued by the construction of a dual system of "micro" and "macro" sociology which in fact gave some sort of order to the distinguishing and the tolerance of the individual theories with each other, yet it brought about no satisfactory solution (Knorr-Cetina/Cicourel, 1981).

Despite these events, even when talking of contemporary sociology one can hardly say that a crisis exists. It goes without saying that this "theory deficit" is regretted in the foreword of diverse anthologies, in speeches or in some articles. Few specialists themselves feel seriously concerned (as long as they do not consider themselves to be theorists). The reason is quite obvious to see: It is quite possible for one to work empirically within his or her specialized field of sociology (hyphen-sociology) and get by perfectly well without "grand" theories.

Furthermore, the almost complete loss of the empirical bases, as happened with the newer theories (Habermas, Luhmann) leads to an increasing alienation of those sociologists working empirically on theories. The quality of their knowledge will be less determined by the inclusion of "grand" theories but rather by another form of empirical work in the same area. Thus theories are being reduced to the level of a contemplative synopsis which one merely has by-the-by and for which not much expense is spared at congresses and symposia. The fact that this "solution" cannot represent a long-term or even a productive solution is proven by the increasing difficulty in research to be able to give some semblance of theoretical order to empirical data.

The unbroken fascination with the classic scientists (Weber, Durkheim, Marx etc.) can be traced back to one simple fact: In their work, a perfectly fruitful integration of rich empirical material with original theoretical thought can be seen. Modern theorists on the other hand seemingly tend to feel that everything which has something to do with empirical material should be consciously avoided.

If sociology intends to maintain its claim to being an integral science, an exit from the "theory cul-de-sac" has to be found. A reconstruction can not however follow in the hitherto way (text studies by sociological and non-sociological theorists). It has to establish the limits of sociology; that is to say that it must by forced from a peripheral position within the science to a more central place within sociological knowledge. However, the process of defining the "limits" can not be led from the traditional -the discussion of theories - "centre" of science. It must be focused on "research practice" where exactly these limits are defined and are continually being relocated. The new and rapidly developed field of research into the sociology of technology lends itself to this attempt.

5 New theoretical potential

One of the most awkward problems relating to the sociology of technology lies in the traditional separation between technology and society. It seems to be attractive to split up the world into a "material" and "social" one as it is used to be done, and to dedicate oneself to the analysis of one or the other of these segments. Yet these attempts continually experience, that the respective other world which was so carefully eliminated for being a "disturbing factor", irrefutably blends back into the picture.

A completely new orientation is required in order to avoid a fixation of social scientific terms upon established technical preconditions. It simply is not enough to research and investigate the development, extension and consequences of technology by using social scientific methods. This does not lead to a comprehensive sociology of technology, but rather to a "commentated" sociology of technology. Thus those protagonists involved in the mechanisation process are able to be named and included according to social characteristics and the different grades of their engagement. However, questions as to the origins and the real relevance of technical progress for the dynamics of human civilisation remain closed to this nearby entrance.

The dictation of technical terms, which is still evident in the compiling of social and technological changes by means of a statistical process, can only be surmounted by the usage of exclusively sociological terms and theoretical concepts. These must be able to question the effectiveness and relevance of technology and, with the same self-evidence, to integrate in just the way like it is in sociology where "sociological matter of facts" (E. Durkheim, 1980) are examined and recorded.

Research relating to the sociology of technology should stand on its own two legs; one should not have to first make arduous detours in order to separate "technology" and "society" before finally one can attempt to filter out the meaningful aspects of technology for society. This is not enough, and has already proven disadvantageous. People who have integrated the most different technologies into their lives mostly without any problems cannot so suddenly begin to do nothing with their respective technology, such a critical and innocent presumption to express new possibilities engenders sociologists reservation. It is obvious that - from the general understanding of a separation between the "technical" and "societal" spheres - even appropriate criticism must be seen as an argument "against technology" and not as a proposal for the "solution to the societal problem". According to this, critical analysis serves thus always as slightly competent interference into an area in which even sociology itself has not been explained completely. Furthermore this is considered as an interference aimed at the hindering of personal legitimate exploiting interests.

A reasonable and generally comprehensible social scientific intervention into the question of technological developments (whose intrusion in societal processes is doubtless) sets out the lifting of the established contradictory ideas of "technology" and "society": "Sociology of technology" should not and cannot investigate "technology" by itself, but rather a technologically structured, "technical" society, and finally that societies structures and features which make technology such a powerful source.

Taking an empirical investigation into attitudes to computers as an example, Nass et al. (1992) show that a "technology-centred entrance" insufficiently enabled people to understand computers. They plead for a radical change to the perspective: In order to examine human behaviour in relation to computers, it is imperative that the machines be considered as machines. The actual astonishing and dubious "nature" of their specific characteristics among societal standpoints can however only insufficiently be understood. The attempt to set in concrete theories on human-machine relationships based directly on the characteristics of machines fails due to the untenable division of technology and society: In the light of such a separation, terms which appear in one sphere simply do not suffice to explain the phenomena occurring in the other. As an alternative to this approach, the American authors formulate the "anthropocentric" proposal which basically deals with human behaviour in respect to their environment regardless whether or not this is determined by other people, machines or computers with human (anthropogenic) abilities/characteristics. According to this claim, theoretical analysis should - as should interpretative models established or uncovered in the research relating to the sociology of technology, which per se do not relate to technologies. In support of Adorno/Horkheimer (1955) and Durkheim (1976) the technical perception of the environment - however explicitly and consciously including "technology" - into sociological research is being established.

The fruitful nature of this action can quite readily be seen in the quoted investigation by Nass et al. (1992) concerning the predictability of acceptance. Technical contact and experience with computers prove the explanation and grounds for the acceptance of new technologies as comparatively surprising insignificantly opposing factors which are also generally relevant for the shaping of societal opinion models and modes of behaviour: Familiarity with foreign cultures and further education make transfers from anthropomorphous characteristics to computers - and as a consequence of which the "acceptance" - much more likely than when these socio-cultural "resources for tolerance" are lacking.

It has already been emphasized that research relating to the sociology of technology must free itself from the division between technical and social systems. If one places the so-called "socio-technical systems" (c.f. Ropohl, 1979) in the limelight, the relationship between technical artefacts and human beings - as part of one integral "system" will have to be newly defined and ordered. These efforts of definition most recently put forward in the objection by Rammert (1989, 146) when it was stated that elements of a system of that kind are heterogeneous <without common characteristics> and the limits to the system are inadequately determinable from outside <arbitrarily>, must be analytically taken into consideration. The fact that this reversion to proven social scientific categories is possible makes Nass et al. with the anthropocentric theory clearer.

In sociology, the more usual terms for social actions in their traditional form may not under analysis grasp technological developments and its consequences - because of an all too common "technocentrical fixation": Due to restrictions of this kind, no consideration of the differences between the system components of "human-machine-interaction" could make any direct or productive use of the term human actions. Nevertheless the transition from a "social" relationship to a "socio-technical" relationship can be understood by introducing a precisely defined term of mechanisation: *>Mechanisation< is transference from social actions on technical apparatus and systems* (Hochgerner, 1990a, 235). In this way, mechanisation in an analogical step begins to have a connection with the elementary sociological term of social actions, for the "intended meaning" (intentional functionalism), and orientation to social subjects, are constitutive (Weber, 1956). What makes mechanisation processes relevant from a sociological perspective is consequently the adoption (incorporation or objectification) of social actions (at least of certain sequences of it), by means of technical artefacts. Functions and achievements of a technical artefact or system take over - more or less completely - the intention connected to an analogical social action, and set off (other) protagonists. Technology can, just like social actions (in Weberian terminology), both passively mediate, as well as actively stimulate subjective opinions on behaviour of often very different subjects (people, groups, institutions).

6 The social implementation of mechanisation processes

Societal relationships can be described according to certain structural characteristics, whereas in particular power, influence and well-being interests have hitherto captured the attention of sociology. But far beyond that the structural elements of human societies extend from the cultural and emotional constituent nature of its members to the material products which they have created. This range can be seen as a *multi-level structure composed of "sediments"* which incorporates the extreme elements of a manifestly material world as well as the latent components of culture (Hochgerner, 1986, 65):

- The extension and meaning(s) of the relevant *mythologies and rituals* in the respective society in question provides the most "internal" and - concerning mechanisation - most resistant sediment of societal structuring.

- In the psychic field, the distribution or different evaluation of *desire and fear* has a structuring effect, thus we may characterize the state of a society in terms of its "societal balance of emotions".

- *Information and communicative structures* play a part essentially - cultural as well as material - in the shaping of social circumstances; this third sediment is at present subject to massive mechanisation.

- *Relationships to power*, dependencies and differences in power, whose existence and change gain expression in political systems and in the distribution of resources; in this sediment mechanisation and technology in various forms play a major role since the initial production of arms and ramparts etc.

- *The material-technical environment* can be seen as simply the most visible surface, even as a kind of "skin" of the society; this top sediment represents the external side of social structures and is as much its product as well as its materially manifestation at any time being.

This theoretical construction of "social structural sediments" discriminates structural components of concrete societies (at a given point in its development in terms of the civilisation) by the degree of their mechanisation and their resistance to further mechanisation. Sediments differ in the way they are mechanisable. In their shaping, the respective *"level of mechanisation"* which has been established can be empirically measured and incorporated into theoretical far-reaching concepts of structure and the development of the society.

The concept of the mechanisation of socio-cultural elements offers an explanation for certain social processes, although the question as to exactly how and why remains open. In a fragile social world, how does the situation arise that in social interactions, complicated, numerously interlinked processes constantly recreate stable development trends which are able to consolidate and even gain long term dominance?

Sociology has been dealing with this basic question since its beginnings. Comte saw the breakthrough from the positive mind which was represented in his three staged law, Marx recognised the relationship between production forces and production relations, Weber found rationalization, Luhmann talkes of autopoiesis, and Beck sees risk as a determinable criteria. However a united belief exists stating that common societal principles do exist which give structure and direction to every day interactions. In every day language they are perceived as "compulsion by object" *(Sachzwang)*, *"Zeitgeist"*, and unavoidable competition, etc.

An original and self-conceived approach is the concept of *"formative principles"* (Hochgerner, 1986). Having grown with history and standing strong in the industrially developed societies, the principles of "hierarchism", "objectivism", and "growth" are at the heart of theoretical frameworks (and in subsequent empirical analysis). However these are *not* signs pretending to show the direction in which development will take, nor do they describe normative expectations referring to the historic dynamics of civilisation processes. Rather, they represent a practical implementation and, in large social systems, a specifically related further development of the micro sociological term "figuration" (Elias, 1978, 142). Included in this are "transformation models" in which people act, according to the situation, "not only using their intellect, but also with their complete body, and their every action in their relationship to each other".

Hierarchism, objectivism, and growth are termed "formative" because they respectively bundle up a range of material and idealistic elements together with organizing social facts (laws, customs and traditions, role expectations etc.) in such a way that the societal formation which is supported by this is in fact secured by their existence. This security of continued existence incorporates remodelling and possible change (throughout the society as well as even concerning its formative principles). For societies such as the industrially developed which recognise "growth" as a constitutive necessity, constant change even guarantees preservation: Such a societal formation can only have continued existence if it is in the position to remodel itself in a controlled way to incessant change. Such regulation, which controls human behaviour according to the specific needs of a certain societal formation given, is termed a >formative principle<. It organises existence and change in social behaviour over several historical eras without itself being restricted to the respective form of that time.

The principle of the hierarchical nature of human forms of organisation has always taken a manifold expression since the "invention of the kingdom" (Mumford, 1977) in e.g. the ancient world, the medieval feudal systems, the Soviet Union just as much as in the western democracies with market oriented economies.

The historically more recent principle of objectivism does not imply priority or exclusive concretization but, above all, an "objectification" and "abstract form" of relationships. Only when human behaviour is standardized, "well educated", and "functioning" - in the technical meaning of the word - will mechanisation (represented by technological instruments or incorporation into socio-technical systems) be accessible to it.

Currently the most relevant and, at the same time the most controversial principle of growth connects increasing economic output with the promise of increasing social security and improved levels of wealth, however increasing demands are also put on achievement and the most efficient form of labour organisation.

Technology is an increasingly more meaningful co-shaping component in the existing social structures and in the ecological and economic conditions of life of the present. Formerly technology primarily took the shape of manifest machines as being material elements of the different principles: Weapons and defence installations above all in the sense of the construction, consolidation - although also the elimination - of social and political hierarchies; an image and symbolic representation of >objective< differences within social behaviour ("technology as culture") for examples in architecture, cars (Barthes, 1982) - yet also in abstract means of exchange like money, "value" in an "objectified" form. Lastly technology became industry, the use of natural resources, increasing production capacities and last, but by no means least, the continual innovation of "new technologies" to a characteristic expression of the growth principle of the time.

The most recent developments claim that the *raw material* information (Weizenbaum, 1990) and the mechanized processing of which lead in the mean time to the fact that the influence of technical systems has long left behind the material level and produces increasing impact on the formation of society even on the levels of ideal values and the leading social norms. Anders (1956) deliberately went further: He diagnosed that "to be means being raw material". However, information technologies and communication technologies are key technologies for the future development of the industrialized societies. Thus we place them at the core of our further analysis.

7 Mechanisation of communication

The most important fundamental assumptions for research into communication stem from a face to face interaction: a person communicates information to another physically existing person by using language or jests. These methods of transferring information can however be extensively transferred - at a rapidly growing rate considering the state of technical research - to a technical system. Existing between the sender and the receiver are one or several pieces of technical equipment; physical presence becomes unnecessary, or even excluded all together.

The introduction of communication technologies and the subsequent phases of stabilization and innovative modification (up to the fax machine and cordless telephone) accompany social change being implemented. The more or less continual process of innovation and the development of new acting systems ensues "inconspicuously". Formulated otherwise: The change from "old" to "new" does not happen abruptly, and the influence on behavioural ways happens by the eclipsing of available orientation models with new ones. The ongoing production of new technology products leads to an objectification and heightens the change of social models in a particularly sustained way.

The interlocking of different technical and psycho-social contexts increases complexity as well as the contingency of inter human relations. Above a certain level however, this improvement has the effect of a "strain", so that a reduction is required, following "substantial rationality" to prevent individuals from getting lost in disorientation. The implementation of the telefax machine for example highlights the trend toward precise operation. In addition to that the time taken to finish a task is shortened and permanent presence in ones business is demanded, which is fulfilled "as best as possible" by the cordless phone.

This evidence proves that the development, extension and the existing modification of new technologies do not in any way represent a social triviality. Such processes exert much more of an influence on the structure of a society and changing social action. The observation that information is becoming a raw material, and the fact that it begins - both economically and sociologically - to surpass the importance of material resources and the production of goods, led to the creation of the fashionable term "information society" (for the first time "Japanese Computer Development Institute", 1972). One of the most creative authors who describes this development not only by using terminology, but also systematically relative to its background and investigates the reasons and prospective consequences is Beniger (1986). He demonstrates that behind the development to the information society stands a fundamental revolutionization of all regulatory processes (those dependent on information), where the current technical innovations are not in fact the cause for, but are themselves the result of existing societal changes which have already been in operation for a long time. "Society is currently experiencing a revolutionary transformation on a global scale. (...) I trace the causes back to the middle and late nineteenth century, to a set of problems - in effect a crisis of control -generated by the industrial revolution in manufacturing and transportation. The response to that crisis, at least in technological innovation and restructuring the economy, (...) amounted to nothing less, (...) than a revolution in societal control" (Beniger, 1986, 6).

This transformation, according to our argument, is visible in a historically hitherto unparalleled wave of mechanisation in the sediment of societal information and communication models. In this way, these models and their functions not only become technologically depicted, but transformed themselves. A concentration of study on this sediment is particularly recommended because the most relevant contemporary events in the industrial and technical highly developed societies can be observed here.

Communication, i.e. essentially the exchange of information, has been shaping the foundations for every active and organised interaction with the environment since the beginning of human civilisation. The social regulation of communication and the need for information is indispensable in order to be able to attain a consciousness over the perceptions and experiences relating to immaterial, as well as material circumstances of inner or outer worlds (Leroi-Gourhan, 1984). Differentiated social roles and status symbols are instrumental structural elements around which complex societies have developed. The subsequent technical infrastructure of the media has brought to light a huge continuing mechanisation (in the theoretical way described above) especially since the extension into everyday lives of the telephone, radio, and other communication technologies. Multiline telephones and language storage systems offer large firms superb possibilities to implement or further develop hierarchical competence and dependence more objectively than before by means of the corresponding programming of the system characteristics.

In that way, a new quality of what has been until now possible technological influence on the communication processes is being achieved. Ever since the existence of *societal memory*, and the possibility to store this, the need to differentiate between message and its storage became immediately apparent. This is, for the development of civilisation, of extreme importance: Social evolution necessarily presupposes that individually determined skills and experiences concerning subjective lives (or the local sphere of influence) can be further transferred and activated under various social circumstances when necessary.

The first forms, knowledge e.g. about assets, climate, ground relations, yearly cycles, were mythological or religious illusions. These were able to support - and to a certain extent still do today - the cultural moulding and corresponding shaping of the society, and to a degree adapt to the respective requirements of that time. The further development of language and in particular writing brought about in addition many abstract symbols which made accessible information in extremely sophisticated and varied forms of report. Nevertheless, even until today, it has hardly been possible to separate written scripts and language from their storage (Young, 1983).

Recently though, telecopiers, satellites, video, translation computers and fax-voice-boxes etc. are becoming more important and are making messages available for many different storage facilities. Thus not only communication networks tighten, there is also a wide range of new ways to multiplex usage of information stored and communicated. These are achievements of digital communication technologies. With this radically new basis, the modern "communication media" are not only able to resolve information from its original creator, but also to process it in a technical way, to generate and to pass it on in a manifold structure. In this way however not all traditional forms of communication are becoming mechanised. In fact this partly occurs, whilst parallel to that a means of communication is born which is different to those sorts of information and personal experience which exclude themselves from mechanisation (Zureik/Hartling, 1987).

The relevance of "mechanisation of communication" as a field of research can be essentially determined in the light of these considerations by stating that the technical development does not stand still at the reproduction of communication or information in an objective way, but shows the effect of formative principles: Communication technologies and information systems are able to, as it has already been pointed out using examples, support and help to develop hierarchy; they objectify processes of exchange in social contexts and against the outer environment; finally they establish a core of one of the most spectacular branches of growth in the global economy. Accordingly technology serves the formative principles and it is not easy to imagine alternatives in technical progress, such as ecologically and socially *appropriate technologies* - unless precisely this *"social environment"* of technology changes.

8 Moulding appropriate technologies by new patterns of their social environment

Political trends, opinion polls, demands on education and other indicators suggest that the principles "growth" and "hierarchism" seem to be under strain or even become weak, whilst the principle "objectivism" still appears rather strong. As a matter of fact the most relevant conclusion to be drawn from our analysis of the social environment of technology is: There must be new patterns concerning the dominance of the principles of formation if the path of mechanisation should become effected. The prospects of appropriate technologies may brighten by remoulding the social environment of existing technologies and new technologies under construction.

Asking for possible key elements and relevant changes we now may return to the role of engineers in this process of changing the social environment of technical progress. Obviously and definitely this role itself is changing both structure and impact. Especially no essential differences exist amongst central ecological and socio-political opinions and technological constitutions which can be characteristically explained directly by the "technological profession". A process has existed since the mid-eighties which solves the particularities of technicians' political persuasions.

These results of our study (Hochgerner, 1992) and similar statistics reveal an eminent potential for the assessment of technology as well as for industrial and environmental policies. The value of these new patterns of opinions was hitherto underestimated and, just as importantly, was not taken seriously by technicians and engineers amongst themselves: Whilst about two thirds of the sample claimed to being environmentally conscious and to having accepted a critical approach with respect to technological development, most of the technicians believe their professional colleagues to be without doubt traditional thinking, undeterred optimists. Yet from those who are professionally active technicians, believe in this traditional concepts amounts to barely thirty percent. Although this figure lies above the average based on the findings of corresponding representative interviewing of a general public, it points towards the fact that even the profession of technicians has developed a clear minority position for the unrestricted concept of progress.

By necessity, every honestly determined practical implementation of opinion into real behaviour and the realisation of alternative technological and political illusions will be opposed by numerous structural and organisational restrictions. Moreover, of course just like in any other profession there are those who simply "move with the flow". Those who are inclined to react to the current trends corresponding to that specific time, perhaps adopt views more critical than they can themselves actively achieve. On the other hand however, a growing potential could be established to move flexible toward the accompanying and corresponding comforts: Technicians, particularly the better qualified and more relevant engineers, show increasing concern regarding the products and consequences of their work (though indifference has traditionally been part of their public image as well as the way in which they saw themselves).

Clearly, representatives of the technological profession are in no way unpolitically minded. Not only do more than half (54 %) of them support the demand that the right to participate should be extended to the employees, a more surprising third of all even proposed an essential qualitative extension in the character of industrial participation. They simply don't want merely to determine themselves with respect to their indisputable own rights and working conditions, but also about professional goals and products: They want to hold a position to be able to make essential decisions directly concerning industrial and environmental politics. In all these areas technologically qualified experts could become partners and active bearers of an offensive political ideal. Such aims ensue from the context of working organisations within industry, particularly with reference to industrial and industry wide further training. Thus a revealed requirement of this specifically qualified profession exists even stronger than before, which has been produced and further developed as a result of their own ability, is able to use ecologically sound controls.

To really do so, the technicians' profession needs possibilities for further development of their conceptions of society and of environmental politics, opportunities within industry for extended participation rights, and industry wide forms of recurrent education. To actually reshape patterns of the social environment of technology it would be indispensable to provide sufficient time during working hours within the industry for discussion and deliberation about work, professional targets, products and product evaluation, as well as the ecological consequences of the developed products. In addition we propose to establish networks of external advisers for working groups of technicians (study groups for engineers and for certain specialized groups etc.) within the industry. As part of the current European transformation situation, a completely new set of research and advisory councils is needed for trade unions, industry as a whole and its workforce. Representatives of the employees, just as employers, nationalized industries and lastly professional representatives of the engineers find themselves in this situation facing a completely new challenge. There is a good chance that technicians (engineers, specialized workers and academics) are ready in an extremely dedicated way to take part. They themselves are the best able to organise and offer technological advice. Social implementations like these could in fact bring about a step in the right direction for ecological change and socially better suited technological development.

References

Adorno, T. W., M. Horkheimer (1955), *Dialektik der Aufklärung*, Amsterdam.

Aichholzer, G. et al. (1991), *Einführung neuer Technologien als sozialer Prozeß.*, Wien

Alemann, U. et al. (1992), *Leitbilder sozialverträglicher Technikgestaltung*, Opladen.

Altvater, E. et al. (1985), *Arbeit 2000. Über die Zukunft der Arbeitsgesellschaft*, Hamburg.

Anders, G. (1956), *Die Antiquiertheit des Menschen.* Band 1: 'Über die Seele im Zeitalter der zweiten industriellen Revolution', München.

Armanski, G. (1990), *Wir Geisterfahrer e.v.*, Bielefeld.

Asdonk, J. et al. (1990), 'Technikgenese im Kontext von Hersteller-Anwender-Beziehungen' in Tschiedel, R. (Hg.), *Die technische Konstruktion der gesellschaftlichen Wirklichkeit: Gestaltungsperspektiven der Techniksoziologie*, München.

Baethge, M., H. Oberbeck (1986), *Die Zukunft der Angestellten. Neue Technologien und berufliche Perspektiven in Büro und Verwaltung*, Frankfurt/M - New York

Bammé, A. et al. (1983), *Maschinen-Menschen, Mensch-Maschinen. Grundrisse einer sozialen Beziehung*, Reinbek

Barthes, R. (1982), *Mythen des Alltags*, Frankfurt/M

Beck, U. (1986), *Risikogesellschaft: Auf dem Weg in eine andere Moderne*, Frankfurt/M

Beniger, J. R. (1986), *The Control Revolution. Technological and Economic Origins of the Information Society*, Cambridge/Mass. - London

Berka, G. (1992a), *Der Aufbruch des Ikarus. Soziologie der Flugtechnik von den Anfängen bis 1914*, Wien.

Berka, G. (1992b), 'Wechselwirkungen zwischen Ideen und Interessen. Beobachtungen am Beispiel der Theorieentwicklung zur Techniksoziologie', in Hochgerner J., S. Katsikides, Hg., *Die Formierung von Interessen*, Wien.

Bijker, W., T. P. Hughes, T. Pinch (1987), *The Social Construction of Technological Systems. New Directions in the Sociology and History of Technology,* Cambridge/Mass. - London.

Braudel, F. (1979), *Civilisation matérielle, économie et capitalisme*, Paris.

Braun, I. (1988), *Computer und Intimität. Der Baby-Computer: Eine techniksoziologische Fallstudie*, Berlin.

Coenen, R. (1985), *Steinkohle - Technikfolgenabschätzung ihres verstärkten Einsatzes in der Bundesrepublik Deutschland*, Berlin - Heidelberg - New York.

Cronberg, T. (1991), *Danisch experiments - social constructions of technology. New social science monographs*, Copenhagen.
Dierkes, M., U. Hoffmann, Eds. (1992), *New Technology at the Outset. Social Forces in the Shaping of Technological Innovations*, Frankfurt/New York.
Dollinger, H. (1972), *Die totale Autogesellschaft.*, Stuttgart.
Durkheim, E. (1976), *Über die Teilung der sozialen Arbeit*, herausgegeben und eingeleitet von N. Luhmann. Frankfurt/M.
Durkheim, E. (1980), *Die Regeln der soziologischen Methode*, herausgegeben und eingeleitet. von R. König. Darmstadt-Neuwied.
Elias, N. (1978), *Was ist Soziologie?*, München.
Garfinkel, H. (1967), *Studies in Ethnomethodology*, Englewood Cliffs.
Gehlen, A. (1957), *Die Seele im technischen Zeitalter. Sozialpsychologische Probleme in der industriellen Gesellschaft*, Reinbek.
Getzinger, G., B. Papousek, Hg. (1987), *Soziale Technik. Antworten auf die Technikkritik*, Wien.
Goffman, E. (1970), *Wir alle spielen Theater*, München.
Grewlich, K. (1986), 'Kooperative Eindämmung möglicher Konflikte in der "Globalen Informationsgesellschaft"', in Kohler-Koch, B., Hg., *Technik und internationale Politik*, Baden-Baden.
Habermas, J. (1968), *Technik und Wissenschaft als »Ideologie«*, Frankfurt/M.
Halfmann, J. (1984), *Die Entstehung der Mikroelektronik. Zur Produktion technischen Fortschritts*, Frankfurt/M.
Hochgerner, J. (1986), *Arbeit und Technik. Einführung in die Techniksoziologie,* Stuttgart.
Hochgerner, J., Hg. (1990a), *Soziale Grenzen des technischen Fortschritts. Vergleiche quer durch Europa*, Wien.
Hochgerner, J. (1990b), 'Die soziale Codierung technischer Systeme', in R. Tschiedel (Hg.), *Die technische Konstruktion der gesellschaftlichen Wirklichkeit*, München.
Hochgerner, J. (1992), *Techniker im technischen Wandel. Analysen zu Veränderungen von Beschäftigung und Qualifikation in technischen Berufen.*, Wien.
Hochgerner J., A. Bammé, Hg. (1989), *Technisierte Kultur. Beiträge zur Soziologie der Technik,* Sonderband 1 der ÖZS, Wien.
Hortleder, G. (1973), *Ingenieure in der Industriegesellschaft*, Frankfurt/M.
Hulsmann, H. (1985), *Die technologische Formation. Oder: Lasset uns Menschen machen*, Berlin.
Japanese Computer Development Institute (1972), *The Plan for Information Society. A National Goal for the Year 2000*, Tokyo.
Janshen, D., H. Rudolph (1987), *Ingenieurinnen. Frauen für die Zukunft*, Berlin.
Jochem, E. (1988), *Technikfolgenabschätzung am Beispiel der Solarenergienutzung*, Frankfurt/M.
Joerges, B. (1988), *Technik im Alltag*, Frankfurt/M.

Jokisch, R. (Hg.) (1982), *Techniksoziologie*, Frankfurt/M.
Kern, H., M. Schumann (1970), *Industriearbeit und Arbeiterbewußtsein*, 2 Bde., Frankfurt/M.
Kern, H., M. Schumann (1984), *Das Ende der Arbeitsteilung? Rationalisierung in der industriellen Produktion. Bestandsaufnahme. Trendbestimmung*, München.
Knie, A. (1991), *Diesel - Karriere einer Technik. Genese und Formierungsprozesse im Motorenbau*, Berlin.
Knorr-Cetina, K., A. Cicourel, Eds. (1981), *Advances in Social Theory and Methodology. Toward an Integration of Micro- and Macro-Sociologies*, London.
Langer, J. (1981), *Ingenieure und Kaufleute. Eine soziologische Untersuchung der technischen und ökonomischen Intelligenz*, Klagenfurt.
Lederer, E. (1938), *Technischer Fortschritt und Arbeitslosigkeit. Eine Untersuchung der Hindernisse des ökonomischen Wachstums*, Frankfurt/M 1981.
Leroi-Gourhan, A. (1984), *Hand und Wort. Die Evolution von Technik, Sprache und Kunst*, Frankfurt/M.
Linde, H. (1972), *Sachdominanz in Sozialstrukturen*, Tübingen.
Linde, H. (1982), 'Soziale Implikationen technischer Geräte, ihrer Entstehung und Verwendung', in R. Jokisch (Hg.), *Techniksoziologie*, Frankfurt/M.
Löwith, K. (1964), 'Das Verhängnis des Fortschritts', in H. Kuhn, F. Wiedmann (Hg.), *Die Philosophie und die Frage nach dem Fortschritt. Verhandlungen des Siebten Deutschen Kongresses für Philosophie*, München.
Maimann, H. (1987), "Glück und Segen früh und spat schenkt Dir Nudleggs Automat". Zur Technisierung des Haushalts, in L. Unterkircher, I. Wagner (Hg.), *Die andere Hälfte der Gesellschaft. Österreichischer Soziologentag 1985*, Wien.
Marcuse, H. (1941), 'Some Social Implications of Modern Technology', in *Zeitschrift für Sozialforschung*, Jg. 9, Nr. 3/41 (pp. 414-439).
Marcuse, H. (1967), *Der eindimensionale Mensch.*, Neuwied - Berlin.
Mickler, O. et al. (1976), *Technik, Arbeitsorganisation und Arbeit*, Frankfurt/M.
Mumford, L. (1977), *Mythos der Maschine. Kultur, Technik und Macht* Frankfurt/M.
Nass, C. et al. (1992), *Anthropocentrism and Computers* (unpublished paper), Stanford.
Nowotny, H. (1979), *Atomenergie. Gefahr oder Notwendigkeit*, Frankfurt/M.
Opp, K.-D. (1970),
Ornetzeder, M. (1992), *Sozial- und umweltverträgliche Technik. Soziologische Dimensionen einer "anderen" Technik am Beispiel der österreichischen Solarkollektor-Selbstbaubewegung*, Wien.
ÖAMTC (1990), *Mobilität in Österreich*, Graz-Wien.
Radkau, J. (1983), *Aufstieg und Krise der deutschen Atomwirtschaft 1945-1975*, Reinbek.

Rammert, W. (1983), *Soziale Dynamik der technischen Entwicklung. Am Beispiel der »science-based industry«*, Opladen.
Rammert, W. (1989), 'Technisierung und Medien in Sozialsystemen - Annäherung an eine soziologische Theorie der Technik', in Weingart, P., Hg., *Technik als sozialer Prozeß*, Frankfurt/M.
Rammert, W. (1991), *Entstehung und Entwicklung der Technik: Stand der Forschung zur Technikgenese in Deutschland.* Veröffentlichungsreihe der Abteilung Organisation und Technikgenese des Forschungsschwerpunkts Technik-Arbeit-Umwelt des WZB, Berlin.
Rammert, W. (1993), *Technik aus soziologischer Perspektive. Forschungsstand. Theorieansätze. Fallbeispiele - Ein Überblick*, Opladen.
Rip, A, (1987), 'Gesellschaftliche Prozesse in der Technikbewertung', in Getzinger G., B. Papousek, Hg., *Soziale Technik*, Wien.
Ropohl, G. (1979), *Eine Systemtheorie der Technik. Zur Grundlegung der Allgemeinen Technologie,* München.
Sachs, W. (1990), *Die Liebe zum Auto*, Reinbek.
Schmoranz, I. (1980), *Makroökonomische Analyse des Informationssektors*, OCG, Wien-München.
Seifried, D. (1990), *Gute Argumente: Verkehr*, München.
Tschiedel, R. (1989), *Sozialverträgliche Technikgestaltung*, Opladen.
Tschiedel, R., Hg. (1990), *Die technische Konstruktion der gesellschaftlichen Wirklichkeit*, München.
Ullrich, O. (1979), *Technik und Herrschaft. Vom Handwerk zur verdinglichten Blockstruktur industrieller Produktion*, Frankfurt/M.
Wagner, E. (1991), *Technik für Frauen. Arbeitszusammenhang, Alltagserfahrungen und Perspektiven der Hausfrauen im Umgang mit technischen Artefakten*, München.
Weber, M. (1956), *Wirtschaft und Gesellschaft*, Tübingen.
Weingart, P., Hg. (1989), *Technik als sozialer Prozeß*, Frankfurt/M.
Weizenbaum, J. (1990), 'Information statt Material', in E. Gröbl-Steinbach (Hg.), *Licht und Schatten. Dimensionen von Technik, Energie und Politik*, Wien - Köln.
Young, P. (1983), *Power of Speech. A History of Standard Telephones and Cables 1883 - 1983*, London.
Zapf, K. (1986), 'Soziale Technikfolgen in den privaten Haushalten', in Glatzer W., R. Becker-Schmidt, Hg., *Haushaltsproduktion und Netzwerkhilfe*, Frankfurt/M - New York.
Zureik, E., D. Hartling (1987), *The Social Context of the New Information and Communication Technologies: A Bibliography*, New York.

8 Innovation patterns

Wolfgang Hofkirchner and Margit Pohl

1 Introduction

In 1991, a study about innovations in Austria was carried out at the Institute for Design and Technology Assessment of the Technical University in Vienna (FLEISSNER et al 1992). The sample was restricted to all technical innovations offered in Austrian periodicals in 1989, although it included both Austrian and foreign innovations. The study was inspired by the work of Alfred Kleinknecht of the University of Amsterdam who invited scientists of several European countries to co-operate in improving the available statistics on innovation output. Not only in Austria is the data relating to innovation output not very comprehensive or precise, because of numerous problems in data collection, and categorisation. On the other hand, the importance of statistically sound information on this topic is self-evident. Therefore, the EC plans to institutionalise research along these lines. There are contacts between our institute and corresponding research institutes of the EC.

The theoretical background for investigation into the innovation process necessitates the crucial question whether or not there are any relations between this process and economic growth. One of the most well-known theories in this context was formulated by SCHUMPETER (1939) on the basis of the long wave theory. In SCHUMPETER's concept, innovations are introduced in a process of "creative destruction". This means that old industries are destroyed while entirely new industries connected with innovations come into existence. The introduction of innovations leads to an economic expansion up to a certain point when the innovative investments are completed and recession sets in. SCHUMPETER's hypotheses were criticised because of their lack of empirical evidence. KLEINKNECHT (1987) tried to overcome this deficiency. He states that, in the light of several recent empirical studies (including his own), the assumptions made in the long wave theory seem rather plausible as well as SCHUMPETER's clustering of radical innovations at certain points in time, at least for developed capitalism (from 1890 onward).

He was able to show (with several restrictions) that clusters of basic innovations coincide with depression periods. This does not seem very plausible if we assume that the risks for radical innovations in these periods are higher than in periods of prosperity. On the other hand the incentives for such projects are higher as well because traditional technologies no longer grant immediate profits. One of the restrictions for this statement is that it is particularly relevant for product innovations while process innovations do not play such an important role in the long wave context. Another restriction concerns the nature of the innovations. KLEINKNECHT distinguishes between basic innovations and "normal" innovations moving along the established paradigms of scientific development. This distinction was introduced by MENSCH (1977). Basic innovations lead to a radical change of existing technology, normal innovations are gradual improvements of products or processes. KLEINKNECHT's results are only valid for basic innovations.

It is evident that precise information on the innovation process is important for the development of the theory of the business cycle. On the other hand it can be valuable for the evaluation of different measures of economic policy. Measurement of the R&D investments of a firm are not always a good indicator for its actual innovative behaviour. The amount of R&D input should always be compared with the innovation output to get an idea of the efficiency of R&D activities. It is still not clear if small and medium sized firms use their R&D input in a more efficient manner than large firms. There may also be differences according to the branch of the firms. The answers to these questions can influence the direction of subsidies for R&D activities.

2 Methodological problems

Although some research has been done recently concerning the innovation process, there are still many methodological questions which remain unanswered. Amongst examples of which are the distinction between product and process innovations, the difference between basic and normal innovations, and the exact date of the introduction of an innovation.

KLEINKNECHT (1987) presented a possible solution for the problem concerning the distinction between product and process innovation. In some cases it is fairly easy to determine the character of an innovation. There is no doubt that a colour TV is a product innovation and a new numerically-controlled machine is a process innovation. However, new apparatus for cancer treatment could be a product innovation for sick people as well as a process innovation for the hospital (for which it is a worthwhile investment). In order to overcome this problem, KLEINKNECHT proposes a more sophisticated scheme of categorisation. He distinguishes between six kinds of innovations:

- 'pure' product innovations

- new medical procedures or apparatus; new drugs

- new investment goods that are primarily aimed at providing new goods and/or services to final consumers (world's first commercial jet aircraft)

- new technical devices, new technological materials with potential uses for a larger variety of purposes (microprocessor, polyethylene)

- scientific instruments

- pure process innovations

There is a simplified version of this scheme consisting only of two categories: 'pure' process innovations, and all other categories (which constitute a very broad product innovation category).
The problem in defining basic innovations is connected with the discussion about the exact point in time when an innovation is introduced. Basic innovations are supposed to be 'great deeds in technology'. They lay the foundations for entirely new industries and are followed by many improvement innovations. But usually it is very difficult to decide which of a number of inventions is the one to constitute an entirely new product or process. The bicycle, for example, was developed between 1818 and 1885. In 1885 the first commercially successful bike was created. At least six inventions were part of this process. It may be argued that 1885 (introduction into the market) was the year of the innovation. In addition to this, we have to distinguish between simple basic innovations (security matches), and highly complex ones (cars, aeroplanes).

3 Research methods

There are different methods to collect data on innovation output. We can roughly distinguish between four of them:

1 Questionnaires

2 Interviews

3 Patent statistics

4 Analysis of periodicals

All these methods have their advantages and disadvantages. The questionnaire-method is rather widespread because it is comparatively easy to perform. But the answers are not very reliable and depend on the person who fills in the form. The interview-method is much more reliable and thorough, but it is extremely expensive. Analysis of patent statistics is reliable in a formal sense but there are inventions which are never patented, and there are many inventions which are never introduced to the market. Innovation counting based on periodicals has three advantages:

- There are no contacts with firms (no rejections to answer which means a representative sample).

- Data collection is cheap because no experts are required.

- Data collection may easily be extended into the past to make analysis of time series possible.

Problems can arise because of the fact that some innovations might not be mentioned in the periodicals, especially process innovations for the own firm. Besides, it might be argued that large enterprises have more possibilities to represent their innovations in periodicals than small or medium sized enterprises. Literature based innovation counting could therefore overestimate the innovative activities of large enterprises. Literature based innovation counting can be advisable because it is at the same time comparatively cheap and the results are fairly reliable.

4 Empirical studies in innovation research

In Austria, there are three different projects which set out to analyse innovations: the information system FINKORD hosted by the Federal Chancellor's office, the information system ÖIB (Österreichische Informationsbörse) supported by the Lower Austrian Chamber of Trade (Niederösterreichische Handelskammer), and the "Technology and Innovation Test" (TIT) carried out by the Austrian Institute for Economic Research (Österreichisches Institut für Wirtschaftsforschung).

FINKORD is not an information system developed for the purpose of research into innovations, but rather a system for the federal co-ordination of subsidies for innovative activities. The aim of ÖIB is the establishment of numerous contacts between developers of new products (scientists, research institutes, inventors,...) and their partners in enterprises looking for new inventions. The third project, TIT, is the only one whose purpose is predominantly the investigation of innovations in Austria. It is a data base containing information about the innovative activity of Austrian enterprises, their size and relevant regional and sectoral parameters. TIT uses the questionnaire method. It was first carried out for the year 1985. A similar research project is now being conducted for the year 1990.

There are several foreign studies on innovations. The most important one for the project conducted at our institute was the one of KLEINKNECHT (1990) who employs a literature based approach. KLEINKNECHT used a very differentiated classification scheme. This classification scheme was also the basis for our study (with slight modifications) to make international comparisons possible.

5 The innovation study at the Institute of Design and Technology Assessment in Vienna

The design of the project

The innovation study at our institute was carried out in the course of a lecture on "Technology Assessment". The aim of this lecture was to impart theoretical and practical knowledge on this subject. The task of the approximately 150 students was to analyse periodicals for the occurrence of innovations. The classification scheme was rather comprehensive and included the following questions:

1 The name of the product and a brief description.

2 Details concerning the source of information (the name of the periodical, the economic sector to which it belongs).

3 Details concerning the Austrian firm who offers the innovation.

4 The degree of complexity of the innovation:

High The innovation is a system consisting ofmany different components (aeroplane)
Medium The innovation is a system consisting ofa small number of components (laser printer)
Low single innovation (new bicycle brakes)

5 The features of the innovation. These features can show the direction of current innovative activity. The results are especially relevant when we compare different points in time, or different countries. We could, for example, check if, in the course of time, user friendliness of new products becomes more important or not. Examples of attributes of the innovations are: cheap, more efficient, has new or better functions, more user friendly, healthier etc.

6 The degree of novelty:

- a completely new or basically changed product or service (bicycle, electronic banking)
- a moderately new product or service (a more energy-efficient machine, better protection for credit cards)
- a new or better auxiliary product (better brakes for bicycles, insurance combined with a credit)
- differentiation of a product or service (a soap with a new perfume)
- process innovation

7 The origin of an innovation: Who developed it?

8 The main market for the innovation

8 Was computer science important for the development of the innovation?

9 The main market for the innovation.

10 Was the innovation developed by a single firm or by means of co-operation between two or more firms?

11 Financing/subsidies for the innovation.

12 Patents.

6 The innovation study at the Institute of Design and Technology Assessment in Vienna

Results

The origin of the innovations Table 1 shows how many innovations were created by the enterprises of a certain sector. The sectors are sorted according to the number of innovations created in this sector. The innovations include Austrian and foreign innovations. The leading sector is electronics (21.62 %). Electronics and electrical engineering together are responsible for about 30 % of all the innovations in the economy.

There are differences between Austrian and foreign innovations. The five most innovative sectors in Austria are (1) electronics, (2) food, (3) machine and steel industry, (4) building and (5) chemical industry. When we consider foreign innovations offered in Austrian periodicals, the order is different: (1)

electronics, (2) trade, (3) chemical industry, (4) electrical engineering, (5) machine and steel industry. Food industry is the second most innovative sector in Austria, but only tenth when it comes to foreign innovations. On the other hand it is evident that foreign innovations are often offered by trading companies.

The potential market for innovations Table 2 shows how many innovations are offered in the periodicals of a certain sector. The main point here is not who produces innovations but who could potentially apply new products or processes. Again the number includes Austrian as well as foreign innovations. The most important potential user is the hotel and restaurant industry (12.25% of all innovations offered) which apparently reflects the great importance tourism has for the Austrian economy. This is presumably also the reason why food industry plays such an important role as a market for new products (9.08% of all innovations are offered in periodicals of this sector). Only by adding the innovations offered in periodicals of the electronics industry and in electrical engineering is this sector able to surpass the hotel and restaurant industry. Innovations which are relevant for environment protection are not very important. Only 0.89% of all innovations are offered in periodicals of this sector. These data are not very encouraging.

The hotel and restaurant industry uses mainly Austrian innovations (116 of 166 offered in the periodicals of the hotel and restaurant industry). Austrian innovations also play a major role in the food sector (70 Austrian innovations are offered in the periodicals of this sector), in the transport and communication sector (59), in the electronic industry (58) and in the machine and steel industry (54). Foreign innovations are offered mainly in the following sectors: electronics (75), iron and metal industry (69), food industry (51), traffic and communication (45), electrical engineering (39).

Complexity, novelty and origin Most of the innovations show either a medium or a low degree of complexity (81.9%):

Table 8.1
Complexity of innovation

Degree of complexity	Number	Percentage
High	226	16.68%
Medium	629	46.42%
Low	481	35.50%
No answer	19	1.40%
Total	1355	100.00%

Only very few of the innovations were process innovations (5.68%). Many of the innovations are only product differentiation, or new or improved parts of a product (54.83%). Only a small part of the innovations consists of completely new or basically changed products (17.49%).

Table 8.2
Novelty value of innovation

Degree of novelty	Number	Percentage
Process innovation	77	5.68%
Completely new/changed	237	17.49%
Moderately changed	278	20.52%
New or changed parts	342	25.24%
Product differentiation	401	29.59%
No answer	20	1.48%
Total	1355	100.00%

If we combine the attributes of the degree of complexity and of the degree of novelty, most of the innovations found in Austrian periodicals were product differentiation with a low degree of complexity.

There are slightly more Austrian innovations offered in periodicals (54%) than foreign innovations, at least as far as innovations could be categorised as either Austrian or foreign.

Table 8.3
Origin of innovation

Origin	Number
Austrian company	56
Foreign company	584
Innovator offers his product	635
No answer	80

If we divide the economy in a primary (agriculture, mining and energy), a secondary (production of goods) and a tertiary (services) sector, the secondary is the most important creator and market for innovations. But the secondary sector produces more innovations than are offered in this sector.

Table 8.4
Sector in which innovation appears

Sector	Innovations produced	Innovations offered
Primary	45	89
Secondary	992	826
Tertiary	318	440

Features The five most important features that innovations (Austrian and foreign) offered in Austrian periodicals are supposed to have are: (1) "user friendliness", (2) "security", (3) "flexibility", (4) "reduces labour cost" and (5) "is cheaper". The pattern is very similar for Austrian and foreign innovations, but there are a few differences. Environment protection plays a more important role in Austrian innovations than in foreign innovations (6th place as compared to the 17th place when we look at foreign innovations). When we consider foreign innovations, speed and precision are more important than in Austrian innovations (5th and 6th place as compared to the 8th and 15th place for Austrian innovations). The least important features are: "changes the modes of human interaction", "saves the cost of transport or storing or capital", "makes new services possible".

We can also make out clusters of attributes if we compute so-called coincidence numbers. A coincidence number reflects the frequency with which two different attributes occur together. There are several clusters of attributes which appear together frequently. There is one group which could be called "economical" consisting of the features "cheaper" and "reduces labour cost" and another one with a similar content consisting of the features "reduces labour cost" and "reduces cost of material". And there is a cluster containing features which are important for users ("security", "more user friendly"). The intensity of coincidence in these clusters is fairly strong. In

40% of all cases, when one of the attributes is mentioned, it goes together with the other attribute.

References

Fleissner, P., Judmaier, P., Schwartz, L. (1992), *Innovationszählung in Österreich*, Wien.

Kleinknecht, A. (1987), *Innovation Patterns in Crisis and Prosperity*, London.

Kleinknecht, A. (1990), *Die Entwicklung brauchbarer Indikatoren zur Messung des Innovationsoutputs* (Manuskript), Amsterdam

Mensch, G. (1977), *Das technologische Patt. Innovationen überwinden die Depression*, Frankfurt.

Schumpeter, J.A. (1939), *Business Cycles: A Theoretical, Historical, and Statistical Analysis of the Capitalist Process.*,2 Bände, New York.

9 Technological innovation in Austrian enterprises: Opportunities and barriers from the entrepreneurs' view

Eva Buchinger

Abstract

Innovation and, more specifically, technological innovation is increasingly being recognized by business executives as both an integral part of business strategy and either a major opportunity for, or strong threat to, their firm´s development. Special attention has to be given to products and new technologies because the long-term competitiveness of a company ultimately depends on the capability of transforming manufacturing core competencies into successful products.

The present paper, on the basis of two different empirical investigations conducted in enterprises in the Austrian province Styria, discusses the opportunities and barriers of the technological innovation processes. The following elaborations are dealing with the entrepreneurs' overall attitudes to "technological innovation", on the one hand, and on "technological product diversification", on the other.

Introduction

The willingness and the ability of business firms to innovate are determinants of the firms´ competitive survival, both in the short run and in the long run. The company which, on the basis of its technological know-how, can produce more goods, produce them more cheaply, overwhelm the market with new products and processes, determine the technological progress and have excellent organizational skills is unequivocally superior (Busch 1987). It should be said restrictively at this point - apart from the readiness to introduce the newest technology in order to secure an advantage in competition (Porter 1985) - the efficient use of the specific process technology and the decision for the right product technology are also important criteria. Technology rivalry among firms and industries invigorates whole economies and provides the main driving force for long-term growth. Joseph Schumpeter (1939) stated the point very dramatically when he wrote that the competition that really matters in a dynamic economy is technological competition, because instead of merely affecting the profit margins of firms it strikes at their very existence.

Nevertheless, most organizations do not find it easy to recognize and accept new ideas, to develop them into viable operational concepts, nor to introduce and commercialize them.

These are also the findings of two empirical investigations, conducted in enterprises in the Austrian province Styria. Both investigations are based on interviews with entrepreneurs.[1] The first investigation, conducted in 1988, comprises the entrepreneurs´ overall attitudes to technological innovation, and the second investigation, conducted in 1991, is an accompanying study to technological product diversification projects in several firms.[2]

Technological innovation and the firm´s competitiveness

Technological innovation can affect a firm´s competitiveness mainly in three ways - cost advantage, differentiation and technological leadership. These three ways can be seen as single strategies, but they are often interconnected, for example when differentiation is founded upon technological leadership.

A firm can gain cost advantage through technological innovation both by product technological change and by process technological change (Porter 1985, Sommerlatte and Walsh 1986). The purpose of technological product development is to reduce cost by lowering material content, facilitating ease of manufacture or simply logistical requirements. Technological process improvement can save costs through learning curve effects such as reducing material usage or lowering labour input. Furthermore, process development enhances economies of scale.

A firm differentiates itself from its competitors when it provides something unique that is valuable to buyers beyond simply offering a low price. Differentiation can potentially arise anywhere in the whole chain or web of a firm and is seen too narrowly in terms of the physical product or marketing practices. As well as advanced servicing techniques and unique product features, sources of technological differentiation for example are proprietary quality assurance equipment, unique production/processes or machines and rapid model introductions (for further details see Porter 1985). Differentiation allows the firm to command a premium price, to sell more of its products at a given price, or to gain equivalent benefits (e.g. greater buyer loyalty during cyclical or seasonal downturns) and is in this way the solution to price competition (Kotler and Armstrong 1990). Differentiation is successful when the supplier activates the potential buyer by use criteria and/or signalling criteria (Porter 1985, Sommerlatte and Walsh 1986). Use criteria might include factors such as product quality, product features, delivery time and applications engineering support. Signalling criteria might

include factors such as advertising, the attractiveness of facilities and reputation.
A firm reaches technological leadership mostly through an extended degree of R&D and with unique technological skills vis-à-vis competitors and rarely through sudden and unexpected technological innovation ideas. Apart from this, it is not desirable in any case to be the technological leader because both technological leadership and technological followership entail advantages and disadvantages. The first mover (the prospective leader) gets the opportunity to define the competitive rules like shaping the way in which a product is defined (e.g. definition of standards), to build up institutional barriers (e.g. patents) and to get favourable access to facilities, inputs or other scarce resources. In some industries the first mover may be in a position to enjoy high profits by selling to buyers who value the new technology highly (Porter 1985, Perillieux 1991). But the first mover often faces disadvantages too because technological development can be very costly and the leader bears the whole risk, especially uncertainty of demand (Sommerlatte and Walsh 1986, Porter 1985). So it can be also favourable to be a follower. The advantages of technological followership lower the cost of the product by learning from the leader's experience, avoid R&D costs through imitation, adapt the product or delivery system more closely to the buyers´ needs by learning from the leader's experience.

Barriers: process versus product innovation

Product and process innovations are inevitably interdependent; both a product line and an associated production process must be taken together as the unit of analysis. Nevertheless, a dynamic model shows the changing character of innovation. Firms (or a fictious productive unit) which are new to a product area will exhibit a fluid pattern of innovation (product change is expected to be rapid); when the usefulness of the new product is better understood, the product change rate is reduced and becomes very low when the product reaches a more or less undifferentiated standard (dominant product design). According to this, major process innovations occur when product use is well-known and a volume significant-enough is achieved (Abernathy and Utterback 1982 and 1978, Utterback 1987).

A result of the accompanying study on technological diversification projects was that the possibility of technological innovation in enterprises is assessed differently concerning product and process. The entrepreneurs state that they like innovations connected with characteristics like low uncertainty, a critical extent of internal know-how and only small to moderate capital expenditure. Low uncertainty means that the entrepreneurs are open-minded to new ideas which find a satisfying basis in the firm's given structure, especially with the engineering, the organizational and the commercial staff. A "good" innovation is, in the words of the entrepreneurs, a *"...feasible innovation..."*. And this in their view is relevant to a greater extent for process innovation than product innovation.

But a "good" innovation in these terms is not necessarily a successful innovation. The latter is characterized by a great demand for a product (as a result of a competitive advantage), achieved through technological innovation. To realize a successful innovation (which means successful marketability), in the opinion of the entrepreneur under investigation acquires a quality which exceeds the notion of feasibility.

This distinction is also seen in a more dramatic way in the attitude study. Table 1 shows that in the entrepreneurs´ view technological product innovation is significantly more difficult than technological process innovation.

This result is confirmed both by Austrian and international studies. While the Austrian "Technology- and Innovation-Test" (Volk 1988) quotes as one of the most important innovation barrier (following missing capital stock) the perceived uncertainty of market behaviour, the result of a study about innovation in small and medium sized enterprises in Austria (Bornett and Neugebauer 1985) is that the greatest problem is market entry. Studies of German, Italian and Norwegian firms show that only few companies pay attention to user needs (Holt, Geschka and Peterlongo 1984); the same is pointed out for the US (Peters and Waterman 1982).

Table 9.1
Attitudes to technological innovation

	complete agreement*	strong agreement
Technological product innovation in my company is practicable today	35.8%	9.4%
Technological process innovation in my company is practicable today	89.1%	16.4%

n= 55

*complete agreement: strong agreement, medium agreement, slight agreement.

As stated above, differentiation through technological innovation is a competitive advantage when it provides something unique that is valuable to buyers, e.g. product quality or image. Here the product is seen as a bundle of attributes that affect the buyers desires and determine the buying decision.[3] The critical point, and so far the point which seems to be a problem for the enterprises under investigation, is to be "close to the customer", close enough to avoid a situation which some of the entrepreneurs under investigation went through. One of the companies, belonging to the electrical machinery industry, is exemplary. The managers identified a problem - to integrate microwave ovens in the kitchen - and developed a solution, a vapour-exhaust system combinable with a microwave oven. They had a fully developed technical version, an attractive design, positive preparatory talks with microwave oven traders and a market-growth in microwave ovens - but the innovation was a failure, and it is not clear till today in which way this failure could have been avoided.[4]

Barriers: risk and uncertainty

Schumpeter (1939) stated that risk bearing is no part of the entrepreneurial function because it is the capitalist who bears the risk. But risk, he further elaborates, nevertheless enters indirectly into the pattern in which entrepreneurs work[5] because every new thing is risky in a sense in which no routine action is. Innovations are always dangerous chances combined with risk and uncertainty. Whereas risk can be estimated,[6] uncertainty implies a lack of predictability concerning the number of perceived alternatives with respect to the occurrence of an event and the relative probability of these alternatives. A further problem is that successful innovations only allow enterprises to earn innovational profits for a certain time because they can soon be forced out of existence by rivals and imitators.

Only a few of the entrepreneurs interviewed by us were consciously concerned with the limitation and estimation of risk. One enterprise for example explicitly executes a risk-return calculation before an innovation process is initiated, another makes a more implicit assessment mainly concerning the question of investment. To cope with risk and uncertainty, different instruments were developed over time like Portfolio-Method[7] or PIMS-System (Profit Impact of Market Strategy) for example[8]. For an overview especially on New-Product-Testing see Henry, Menasco and Takada (1989).

But most of the enterprises in both investigations tend to avoid the conscious pursuit of risk. They prefer innovation with low investments *"...no dependency on banks..."*, and realizing the project *"...by drawing on one´s own resources within one fiscal year..."*. Asked for measures they would like to claim, a number of entrepreneurs called (among other things) for tax exemption, general government aid and firm-specific subsidies, protectionist measures, and one demanded governmental technology and market monitoring.

Overcoming barriers: organizational structure

As mentioned above, enterprises do not find it easy to recognize and accept new technological ideas, to develop them into viable operational concepts, to introduce and commercialize them, because most of them are bureaucratic[9] organizations and such organizations are created to handle routine tasks and to lend stability to human relationships to gain organizational efficiency (Perrow 1986). To ensure that such organzations can nevertheless cope with innovation, specific criteria like decentralization, complexity, low formalization, interconnectedness, "slack", autonomy and, last but not least, system openness (Rogers 1983, Souder 1983, Peters and Waterman 1982) have to exist beside the bureaucratic structure. A cautious trade-off process is needed here.

Table 9.2
Attitudes to technological innovation

	complete agreement*	strong agreement
Technological innovation demands activity - responding is not enough n= 55	98.2%	58.2%

* complete agreement : strong agreement, medium agreement, slight agreement

Some of the entrepreneurs of the accompanying study reported that innovation in their company is naturally the task of the development-, production- or construction-head. Thus, they leave innovation dynamics inside the traditional functional organization structure. But most of the interview partners were conscious of the necessity to overcome both the inflexible functional and hierarchical circumstances. One statement, for example, was: *"...the person had to be liberated from the organizational structure, ideally by one hundred percent, but at least to a degree of fifty percent. ... The top management must set him free and provide him with resources and staff."* This is in fact an attempt to harmonize the innovative and steady-state needs of a firm[10] and to overcome the so-called "organizational dilemma" (Wilson 1972) in this way. To organize this is probably easier in the investigated small and medium sized firms than in big companies (Peters and Waterman 1982, The Economist 1992, Drucker 1985, Schuhmacher 1973). The attitude study confirms the necessity of an initiating top management as shown in table 2.

Overcoming barriers: motivation

"Treat people as adults. Treat them as partners; treat them with dignity; treat them with respect. Treat them - not capital spending and automation - as the primary source of productivity gains." These are fundamental lessons from the excellent companies research by Peters and Waterman (1982 p.238). This respect for individuals has to be seen in a profit-oriented framework and should ensure interest to achieve "extraordinary results" through "ordinary people". The statement further expresses that in the information-based organization[11] individual responsibility and self-discipline, which are required in such complex structures, are not reachable alone through motivation strategies like quality circles or monetary incentives.

The diversification study shows that motivation is, at least in part, a known and applied instrument and effects mainly the two dimensions "appreciation" and "information". Motivation happens when the top management starts the innovation initiative and pursues its positive execution, in this way the appreciation will be transmitted to the innovation-project-people and these *"...will be appraised by their colleagues; perceptible attention and approval arise."* Experienced entrepreneurs emphasize the early information-flow to the whole staff and the dialogue especially with the opinion leaders. Altogether an innovation has to be promoted.[12] The attitude study (Table 3) confirms this statement, expressed in a high assessment of staff innovativeness by entrepreneurs.

Table 9.3
Attitudes to technological innovation

	complete agreement*	strong agreement
My professional colleagues are competent and ready to execute technological innovation today	92.7%	14.5%

n= 55

* complete agreement: strong agreement, medium agreement, slight agreement

While Schumpeter concentrates on the entrepreneur´s role, whose personality is characterized above all by an innovative attitude, the recent human relations model places attention on staff members (Perrow 1972, Gussmann 1988, Corsten 1989) and emphasizes both the importance of leadership and the total organizational climate seen as corporate culture. Whereas the latter expresses the shared attitudes, values and beliefs, the management task is one of "transforming leadership", in which the leader promotes and/or protects desirable values such as participation, networking and autonomy (Peters and Waterman 1982, Katz 1964, Vroom 1964, Bleicher 1974).

Conclusions

The technological innovation process inside a specific enterprise shows itself to be a "fuzzy phenomenon", has a lot of varied dimensions and is perceived and explained in a different way each acting entrepreneur. Nevertheless, on the basis of the results stemming from the two empirical investigations, it can be justified to emphasize some findings concerning both innovation barriers and opportunities to overcome barriers. The innovation barrier relevant to product innovation can be seen as "hard fact", because competition is very strong in some industries, especially in degenerating markets, and it is not really possible to predict which product features the buyer will prefer in future. Uncertainty, risk and corporate culture (organization, motivation) belong more on the border of the "fuzzy phenomenon". The generation of innovative cultural values and the establishment of creative climates, however are the management tasks of the future; therefore the roles of entrepreneurs and managers are changing.

Notes

1. Entrepreneur means the firm´s proprietor or managing director.
2. The investigation on attitudes to technological innovation covers 55 firms. The investigation on technological product diversification is based on six projects accomplished between 1988 and 1991. All enterprises under study are small and medium firms and belong to different industries. For detailed information see Buchinger (1989 and 1991).
3. For an overview on the difference between the buyer behavior of consumers and of organizations (industrial buyers, reseller buyers, government buyers) see Kotler and Armstrong (1990).
4. For the problem of producer-consumer-interaction see the extensive marketing publications, for involving the user in the technological product innovation process see for example Holt (1988) or Rothwell and Gardiner (1985).
5. i) Entrepreneur and capitalist in one person; ii) qua entrepreneur he loses other people`s money; iii) riskiness makes it more difficult to obtain the necessary capital.
6. Investors are averse to risky situations; when it is impossible to insure or diversify the risks, a risk premium (normal return for riskless investments versus extra return for highly risky investments) added to returns is required to offset the investors risk aversion (Samuelson and Nordhaus 1989).
7. For a clear description related to the life cycle-concept see the practitioners Streicher and Turnheim (1987).
8. For an overview see Buzzell and Bradley (1987).
9. Rational-legal bureaucracy based on rational principles (Weber 1922); rational in terms of management interests.
10. Autonomy and flexibility for innovative activities (innovation development) and clearly defined functional structures for innovation implementation and routine problem handling.
11. This is a knowledge-based organization, composed largely of specialists

(knowledge lie primarily at the bottom in the mind of these specialists) who mainly direct and discipline their own performance; rewards, recognition and career opportunities have to be developed, and a unified vision must be created (Drucker 1989).

12. For the concept of "power promotor" and "technical promotor" see Witte (1973).

References

Abernathy, William J. and James M. Utterback (1982), 'Patterns of industrial innovation' in William L. Moore and Michael L. Tushman (eds.) *Readings in the management of innovation*, Pitman, Boston

Abernathy, William J. and James M. Utterback (1978) 'Patterns of innovation in technology', *Technology Review* ,June-July

Bleicher, K. (1974), 'Führungsstile, Führungsformen und Organisationsformen', in E. Grochla (Hg.) *Managment - Aufgaben und Instrumente*, Düsseldorf

Bornett, Walter und Herbert Neugebauer (1985), *Innovationshemmnisse in Klein- und Mittelbetrieben*, Signum, Wien

Buchinger, Eva (1991), 'Betriebliche Innovation und Anforderungen an den Innovationsberater aus der Sicht der Unternehmer', in Josef Fröhlich et al. *Entwicklung und Testen eines Modells für technologieorientierte Regionalentwicklung durch Innovationsförderung unter besonderer Berücksichtigung der Obersteiermark*, ÖFZS Nr.2167

Buchinger, Eva (1989), *Die Einstellung von Unternehmensleitern zur technologischen Innovation unter Berücksichtigung des technologischen Standes ihrer Unternehmen*, ÖFZS Nr.4494

Busch, Rainer (1987), *Internationale technologiebestimmte Wettbewerbsfähigkeit und Forschungs- und Technologiepolitik*, Peter Lang, Frankfurt

Buzzell, Robert D. and Bradley T. Gale (1987), *The PIMS principles - linking strategy to performance*, The Free Press, New York

Drucker, Peter F. (1989), *The new realities,* Harper&Row, New York

Drucker, Peter F. (1985), *Innovation and entrepreneurship*, Harper&Row, New York

Henry, Walter/Michael Menasco and Hirokazu Takada (1989), *New-product development and testing*, Lexington, Massachusetts

Holt, Knut (1988), 'The role of the user in product innovation' in Gerhard Rosegger (ed.)(1991), *Managment of technological change*, Elesevier, Oxford

Holt, K./ H. Geschka and G. Peterlongo (1984), *Need assessment - a key to user-oriented product innovation*, Wiley, Chichester

Katz, Daniel (1964), 'The motivational basis of organizational behaviour', in Victor H. Vroom and Edward L. Deci (eds.) *Managment and Motivation.*,

Penguin, 1970
Kotler, Philip and Gary Armstrong (1990), *Marketing*, Prentice Hall, Englewood Cliffs
Peters, Thomas J. and Robert H. Waterman (1982), *In search of excellence*, Harper Collins, New York
Perillieux, René (1991), 'Strategisches Timing von F&E und Markteintritt bei innovativen Produkten' in BOOZ, ALLEN & HAMILTON (Hg.) *Integriertes Technologie- und Innovationsmanagement*, Schmidt, Berlin
Perrow, Charles (1972), *Complex organizations*, McGraw-Hill, New York
Porter, Michael E. (1985), *Competitive advantage*, The Free Press, New York
Rothwell, R. and P. Gardiner (1985), 'Invention, innovation, re-innovation and the role of the user: a case study of British Hovercraft development', in Gerhard Rosegger (ed.)(1991), *Managment of technological change*, Elesevier, Oxford
Samuelson, Paul A. and William D. Nordhaus (1989 13th ed.), *Economics*, McGraw-Hill, New York
Schuhmacher, E.F. (1973), *Small is beautiful*, Abacus
Schumpeter, Joseph A. (1939), *Business cycles*, Porcubine Press, Philadelphia (1989 reprinted)
Sommerlatte, Tom and S.Ian Walsh (1986), 'Das strategische Management von Technologie' in Armin Töpfer und Heik Afheldt (Hg.) *Praxis der strategischen Unternehmensplanung*, Poller, Stuttgart
Streicher, Rudolf und Georg Turnheim (1987), *Strategisch planen - managen*, Manz, Wien
THE ECONOMIST (1992), 'Microsoft - top of the world: Can anything stop its phenomenal growth?', April 4th.
Utterback, James M. (1987), 'Innovation and industrial evolution in manufacturing industries', in Bruce R. Guile and Harvey Brooks (eds.) *Technology and global industry. National Academy Press*, Washington
Volk, Ewald (1988), *Die Innovationstätigkeit der österreichischen Industrie*, WIFO, Wien
Vroom, Victor H. (1964), *Work and motivation*, Wiley, New York
Weber, Max (1922) *Wirtschaft und Gesellschaft.*, Mohr, Tübingen (1985 5.A.)
Wilson, James Q. (1972), 'Innovation in organization', in John M. Thomas and Warren G. Bennis (eds.) *The management of change and conflict*, Harmondsworth
Witte, Eberhard (1973), *Organisation für Innovationsentscheidungen*, Otto Schwarz & Co, Göttingen

Transformation of Patterns III

SOCIAL, ECONOMIC AND POLITICAL SYSTEMS

10 The role of industry in national research systems. The case of Hungary, CSFR and Austria

Ulrike Felt

Under the heading "research system in transition" the changes observable at several levels of the research system are being discussed by numerous analysts of the present day situation. They identify global trends such as internationalization and sharing of facilities, of emerging stronger links between universities - or academia in general - and industry as well as a growing interest in what one could label most suitable "strategic science".[1] In particular the relation between industry and university and its influence on the science system as well as the role of scientific research in the economic development of a nation-state have advanced to the forefront of the science policy debate of the 1990s. Research and Development (R&D) more and more is seen as vital to the economic success of firms and regions with universities benefiting from this in terms of investment, and it seems as if the importance of applied research, diffusion of results and higher education is still growing in industrialized countries. Is this a definite change in the structure of the R&D networks, are we witnessing the negociation of a new contract binding science and society at the end of the 20th century or are we maybe simply watching a passing wave?

To start with it is important to understand why this relationship of science and industry has become so crucial for the further development of the national science systems. Indeed when looking back in time we can surely trace back many of the themes prevailing in the actual discussion. "Public pronoucements on the desirability of links between industry and academe go as far back as to Francis Bacon. Or the discussion of "evaluation" or the key-word "selectivity" in one form or another have been key-factors in the building of scientific institutions and scientific carreers since research went professional in the 19th century".[2] But there is nevertheless one rather fundamental difference in the present-day situation: these aspects were so far not regarded as indispensable tools of science "management". Now a well functioning interface of the science system with the industrial sphere belongs to the key-aims when restructuring national R&D networks. Of course the science system has always been exposed as well as been responsive to political, economic, military and industrial forces however it seems - with a different degree of intensity.

Moreover this discussion on the transition of the science system has to be seen in the framework of a wider discussion about whether or not science has entered a "steady state" and what consequences this would have on the strategies when arguing for funding as well as on the whole R&D system. Indeed when governmental science budgets stagnated or even reduced in real terms industry appeared as an alternative financial partner guaranteeing a long-term financial survival. However it is important to pose the question whether the financial ressources coming from the side of industry should be seen as a short term replacement of other financial sources, if they will continue to play a central role in the funding of universities or if the government has definitely left the role of the main supporter of universities. What will this mean on a long-term for the education as well as for the research? Alongside with the changing funding mechanisms the influence on the intellectual organization of research should not be underestimated. This concern was expressed in many discussions and articles in science policy studies and the three empirical cases which are presented there show rather clearly in how far the linkage of the research domain to a weak industrial base can turn out to be quite disastrous. Will the push toward applied research exerted by industry lead to the neglect of basic research or if the latter is to survive in this financially tight situation what are the arguments for the economic usefulness of basic science ?[3] However one thing seems to be clear: basic research has definitely lost the exceptional position it held for many decades and is more and more forced to accountability and justification of investments.

Henry Etzkowitz has labelled this change in the industry/academia relationship we are observing "the second academic revolution".[4] While in the 19th century the first academic revolution took place, meaning that research obtained a definite place inside universities, the "second academic revolution" consists in establishing the task of economic development alongside with research and teaching inside the universities. The aim is thus to restructure the institutions in such a way to give place to an integration of these three factors. This would however mean, that universities would be bound closer than ever with the sphere of national economy theys would partly develop entrepreneura character and thus would come to play a completely new role.

However there is an additional rather interesting feature in the discussion around "science systems in transition". While it is commonly assumed that all these developments are an outcome of and are steered by national science policies, Arie Rip puts forward a rather different and indeed interesting hypotheses. He far more tends to see science policy as an ex-post rationalization and he tries to interpret the observed trends far more as an outcome of diverse influences on various levels of the system. This hypotheses was strengthened by empirical investigations on industry/university relationships in the U.S. Well established cooperations between these two spheres - as it seems - can best be interpreted by a collusion of interests and a coincidence of a variety of factors, ranging from interest on the national level, to the wish for a strong industrial research partner, from personal relations to the hope to gain additional financial ressources. Thus it is less the set of policy regulations elaborated on the political level which decides on whether or not alliences are formed than the general climat prevailing inside the system. These empirical conclusions seem to be valid for the US science system, however it is of interest to discuss up to what degree and under which general conditions such findings can be regarded as valid for the three countries investigated here.

Thus in what follows I will start with a short historical sketch of the development of the respective national research systems under particular consideration of the role of industry played in it. Along with the argument of Rip it will not only be of interest to look at the science policies as they were evolving over time, or to look at the institutional changes and the way networks were built and rebuilt. Far more it is central to identify the different influences - internal as well as external - exerted on the science system, to look at the way alliences were built up, to investigate discrepancies between formal and informal structures which collectively shaped the present situation. This allows us on the one hand to understand the present-day discussion on the reform of the science systems and on the other hand to identify the constraints which are inherent to the national science system.

Why comparing the three countries? Indeed they appear rather different on a first look, but when investigating them closer some interesting parallels emerge. All three societies are in a phase of transition from the dominance of politics and bureaucracy to a democratic system in which economy, technology and science should belong to the key-issues. The model for the change in all three countries is a strongly market oriented model, i.e. international competitiveness is at the core of the discussion. Having western industrialized nations as a model education, knowledge production as well as knowledge transfer are expected to be structured more efficiently in future. And in the past as well as in all future scenarios industry plays a particularily important role.

Finally, when dealing with these three countries we have to be aware of the difficulties involved. The theme is rather broad and the "observed object" rather complex, the systems still undergo rather quick changes and the process of restructuring is far from being terminated. We therefore are forced to stay on a macro-level, trying to identify fundamental changes and problems with respect to the questions posed. In the conclusions we will then try to draw attention to the advantages but also to the dangers of building too tight relationships between science and its economic environment.

Industry - Academia relationships in Hungary, the former CSFR and Austria. Some historical perspectives

Hungary [5]

From 1949 onwards the Hungarian science system was organized according to the "Soviet model" meaning a strict separation of basic science, higher education and application oriented research. The science system became increasingly fragmented while at the same time decision making as well as financing were being handled centrally. The universities (as well as the *Hochschulen*) which had followed so far the German university model, i.e. unification of research and teaching, now were restricted to be mere teaching institutions. Not only teaching was being separated from research but also basic and applied research were separated from each other institutionally, the former being concentrated in the National Academies of Science and the latter in institutes under the auspices of different ministries. The Academies of Science were enlarged and restructured for this purpose and also became the key-task of graduate education and the authority to issue academic degrees. Both domains - basic and applied research - were again completely decoupled not only from the universities but also from the enterprises which could be the potential users of the research results produced.

The reasons for this clear separation of research and teaching were both of ideological as well as of functional nature. On the one hand it was believed that by concentrating intellectual as well as infrastructural means and by releasing researchers from their teaching charge they would better master the increasing specialization in the sciences. However the negative long-term consequences on the education of students were overlooked completely. The ideological background on the other hand, was to exercise state control of research and to assure the protection of the young students from the "ideologically dangerous innovative specialists".[6]

Although from 1957 onwards several actions were taken to loosen this strict separation between research and teaching, these efforts failed mainly because of infrastructural and financial problems. Within the universities there was no adequate equipment to do good research, not enough space and the personnel was badly paid. This is why the monopoly of research stayed with the Academy and the external research labs and also the attribution of academic degrees remained a priviledge of the Academy.

With the introduction of the economic reform in 1968 the strict central planning was replaced by a market oriented economic policy using the market stimulus for orienting the companies. Linked to that also the role of scientific research was redefined, being now focused on practical applications and on strengthening the relationship between research and economy. Like that one hoped to overcome most efficiently the prevaling economic crises. Scientists were thus not to follow their intuition and their research interests but were expected to produce results for and enter contracts with national enterprises. The logic was even pushed further by freezing the basic research budgets - additional money needed had to be gained with contract research for industry.

While the scenario could have worked out theoretically the economic reform failed in its major features. Real competition never occured and on the domestic market price and quality played an inferior role. The economy of shortage could never be eliminated completely. However science had lost with this reform its special financial status within the Hungarian system and was from then onwards closely bound to an inefficient industrial system. The situation is well summarized in a World Bank report:

"Ready CMEA market for relatively low quality products, and heavy protection for industries both from foreign and domestic competition and from the threat of bankruptcy resulted in obsolate products, shoddy production methods, outdated infrastructure and public services, technological stagnation on the shop flor and environment degradation".[7]

Indeed the reform of the science system based on the assumption that there was industry with interest in innovation and input from the side of research. This, however, turned out rather quickly to be completely erroneous. Quite on the contrary there were several reasons for the Hungarian industry not to be interested in innovation: Different to the situation in market economies, the managers were not in the position to have the necessary basic resources of the wished quality and the quantity needed to fabricate the products. It was virtully impossible to decide for or against a specific product - one had to take what one could get. Linked to that one was confronted with three main problems:

- The necessary ressources for innovation were not available.
- The innovation did not suit in the existing production process and was therefore not "needed".
- The different enterprises cooperate vertically and thus the gain of innovation can easily be destroyed on the next following levels.

Innovations would have to follow the rigid, inefficient and outdated production structures and new procedures were not really wellcomed. However, although there was no demand for innovation the Hungarian enterprises were forced to build up financial reserves and to spend them on technological development.

The decreasing basic financing of non-industrial research labs forced researchers to look for additional money. Thus these funds from industry became an important mean to survive. To understand the importance of this ressource a closer look at the the financial situation of research institutes and university institutes is needed. The former were financed with less than 50% by money coming from the state budget (in the case of academy institutes it is 40%) and the rest had to be found mainly by contract research. Similar things can be said about the university institutes. Also for them nearly half of the budget had to be gained by contract research.[8] But this dependence on the revenues from industry together with the bad economic situation had a rather negative influence. Industry was oriented towards basic applications without wanting a high degree of innovation. Consequently the R&D sector became a confunsing field of research, experimental developments, technological developments, production and marketing.[9] But it also meant that basic research got increasingly negelected. In the late 80s only about 10% of the total amount on R&D expenditure went into basic research while the average for industrialized countries laid at 15-19%.[10]

What is more application oriented scientific research was linked to the low technological level of hungarian industry and therefore most of the contracts dealt with routine testing of material, quality checks, and other tasks which enterprises liked to contract outside.[11] There were only few exceptions such as agricultural and biological research which was quite successful. In most of the cases however the collaboration between industry and academia was more a community of interest, enforced by the fragility of the financing mechanisms.

A consequence of this unsecure financial situation and the need to do contract research was a mixing of scientific standards with economic needs. Katalyn Balázs formulated this rather clearly: "The original rigid differentiation of the types of institutes and types of research ceased, along with the institutional hierarchy and disunity of the innovation chain - that had been the basic idea behind the set-up of the R&D network. Now a sort of cooperative, homogenizing tendency can be observed with regard to the different kind of research."[12] As the enterprises failed due to the internal logic of the economic systems to have innovative activities, scientists themselves began to do development work.

The former CSFR[13]

The Czechoslovakian situation showed in a certain sense quite some similarities with the Hungarian case, even though the difficulties are more accentuated. When in 1948 Czechoslovakia became part of the eastern block it also took over centrally planned economy and with it also the R&D network was reorganized according to these ideas. Administrated centrally the scientific and technological development was coupled strongly to the economic development. Following the "Soviet model" in 1952 and 1953 the Czechoslovakian and the Slovak Akademies of Sciences were founded and a whole network of research institutions was set up. This meant that quite to the contrary with other western countries but similar to Hungary also here the Academies play an important place in the basic research area. What is more the Academy also played a central conceptualizing role in the R&D system. It developped research plans which were then submitted to the government and the state commissions, it had a kind of control function for the basic research carried out in other research units and it was responsible for attributing academic degrees. It was thus not only an important scientific institution, but also had a rather strong political influence.

Applied research, under the auspicies of Ministries and the Central Planning Commission, was carried out either inside industry or within special independent research institutions. These research instutes had the central position in the R&D network. More than 2/3 of the total R&D personnell was working in industrial research labs and obtained 65% of the funds invested.[14] This strong bias towards applied research had an obviously negative consequence for basic research. Universities completely lost their status as research institutions. Similar to the Hungarian case they were left with the teaching obligations. Bad infrastructure, no incentives from outside and the enormous teaching load led to the fact that slowly but steadily the relation between research and teaching became completely destabilized.

The fourties and fifties were dominated by the idea of fast expansion of the science system however without forseeing the necessary organizational structures. It did not take very long until the problems of central steering became rather visible in the sixties. Research and teaching were coupled in a rather inefficient way and the centrally developped plans for industrial research could not be fullfilled. As a solution to this problem one tried to create some free space for the researchers to allow for creativity, but already shortly after, in the beginning of 1968 all these tendencies of liberation were cut off abruptly. Bureaucratic steering was again at the center of science policy in the seventies.

With the spring of Prag all efforts to have elements of market economy incorporated in the science system as well as in the economic policies ended. Central planning was <u>the</u> political concept and the declared aim was to foster innovation and technological progress. One seemed to be convinced that by having a strict planning concept one could work more efficient and more rational. Like that structural changes would become visible earlier and it would be possible to use this advantage.

The political influence of science was enormous. In particular after the clearings of 1948 and 1968 there was a wave of immigration to the west of scientists which were not conform with the political system. Many of the important posts inside the R&D networks were occupied by system conform scientists who in many cases were not scientifically qualified persons. But also the scientific cooperation suffered largely from the political situation. Cooperation concentrated for political reasons on the eastern countries, and in particular on the Soviet Union. Like that it was possible for a long while to keep up the belief that the nation belonged to the leading ones in science and technology. The isolation as a consequence was far stronger than this had been the case for example in Hungary, a fact which showed up rather clearly after the "velvet revolution": Western experts knew hardly on the state-of-the-art in Czechoslovakian science. Also the setting of priorities inside the science system was made along criteria external to the science system, namely the fullfilling of the plan. What is more the state plan was made by burocrats who often were not qualified for their task. This was accentuated by prevailing corruption, lacking performance and sense of responsibility. Finally, as the "market" was oriented mainly towards the Soviet Union innovation was not really needed - also outdated products could be sold easily.

Similar to the Hungarian case this close coupling of research and industry had rather negativ effects. There was no pressure for innovation from industry and research institutes were mainly paid to solve small technological problems. Without pressure from outside the system remained static. And even though the financial ressources increased the output was poor. However the ressources were big enough to trigger no internal rivalries for money: there was a kind of social contract between industrial authority and research autonomy; Money was given in exchange for loyality.

The Czechoslovakian case is still a bit further complicated by the fact that research institutes had been turned into state enterprises in 1988 and now it has to be decided what should be done with those structures in times of privatization.

Austria[15]

Regarding the Austrian science system in comparison with Hungary and Czechoslovakia the main characteristic is the dominant position of the universities. This observation can easily be documented by figures on distribution of personnel and financial resources. Universities employ more than half of the total R&D staff and spend more than one third of the global R&D budget. Both figures deviate from the international average. Generally more than 50% of the staff is working in industry while only about 15% of the budget is spent by universities (OECD average).

The central position of the universities as well as their quasi monopoly in teaching has the direct consequence that problems in the universities touch in a dramatic way the whole science system. This explains why most of the discussions in Austria turn around the universities and their reform. Austrian universities were since the reform in the mid-19th century oriented by the model of the Humboldt university. Professors became public servants and possessed within the universities unrestricted power. The administrative power, however, remained with the ministry. This system remained virtually unchanged until 1975 when a new university law came into force which foresaw more democratic decision making structures. Concerning the budget, personal decisions as well as basic educational decisions the ministry kept its central position.

Looking back to the period since World War II it is surely correct to say that until 1960 research was of virtually no importance in Austria. One reason for this was the structurally weak industry which was just being built up and which mainly relied on the import of technologies and had a very low degree of innovation. First measures to improve the situation were taken in the late 60ies, with key-events being the creation of two research funds - one for basic and the other for applied research - as well as the establishing of a Ministry for Science and Research in 1970. For the first time science and technology were identified as potential key-solutions for an economic situation which became increasingly difficult. This institutional changes had to be seen in the context of larger efforts of the new socialist government to increase the number of people with higher education. The elitarian image of the university should be given up and university education should become accessible to everybody.

The main reason for the present problems of the science system goes back to the seventies and eighties, i.e. to the enormous expansion of the educational system. Not only the increasing costs of such a huge institution turned out to be difficult to cover, but the main problem lays in the fact that the old structures stayed and were not really adapted to the new challanges. What is more the Austrian university system is based on the individual autonomy of the researcher, which had on the system level rather negative consequences, as the social and organizational structure offered little incentives and possibilities to collaborate across institutional boundaries.

With raising financial needs also the pressure for control of the ressources put at scientists´ disposition raised. Both on the political as well as on the university level accountability was in demand. The reaction to this pressure was an increasing bureaucratization and centralization of competence withing the ministry.

While all these structural difficulties are of importance, the main problem surely was and is the continuous underfinancement of Austrian science. Compared to other industrially developed nations which spend about 2.5% of their gross domestic product on R&D, the value for Austria is only about 1.5%. And this value has only been reached after great efforts made in recent years. While most of the money went into a quantitative extension of the system one overlooked the most crucial infrastructural needs, leading to a dramatic lack both in terms of space as well as equipment.

However another central problem is the extremely weak collaboration of industry and universities. Industries - mainly small and medium sized enterprises - had a low degree of innovation and thus were not very interesting as partners for universities. The universities on the other hand were caught in their traditional images and insisisted to see basic research as their main task. Applied research was only regarded of secondary importance. What is more for quite a while money from industry was not regarded as indispensible. Therefore although the importance of the role of a well functioning university/industry relationship was realized and measures were taken to foster applied research and to facilitate knowledge transfer - there were nevertheless major problems in turning scientific knowledge into viable technological products which could be sold on the market.

Even though there was awareness about the importance of industry in the national R&D system it took another ten years until a more realistic technology policy was formulated. It was not part of science policy but had its own structure and hopefully also would develop its own dynamics. The aim was to bring universities closer to industry and vice verse and to create a stable and regular basis for interactions. Expression for this new understanding of research policy, industrial and innovations policies was the concept for technology policy in 1989.

Despite all efforts the fact remained that Austrian industries invest far too little in development and implementation of new technologies. Partly this can be explained by the small size of the enterprises, but even bigger concerns such as Austrian Industries invest far less than their international counterparts. What is more, most of the investment flows into the development and far too little into basic research.

Further there are quite clear lacks in the organizational structure. The competences in the technology sector are divided among several ministries and coordination does not work efficiently. What is more, it was stated clearly in an OECD report that democratic decision making structures are completely missing in this domain. Decisions are taken in very small groups of people within ministries, consulting teams are not really independent and initiatives in this sector mainly come from the ministries. What is more these structures are rather complicated and heavy.

For close with one could say, that even though there are quite some efforts undertaken to render the research system more efficient it seems that Austrian universities are far away from entering the "second academic revolution".

Concluding remarks

What becomes clear from the accounts given above is the fact that the R&D systems in each of the three countries have to be described as single, losely articulated but essentially indivisible units, where very diverse activities, ranging from basic research to near-market technological development, interpenetrate and combine within a variety of institutions. In all three countries a high degree of dissatisfaction with the efficiency of the science system was reached and the call for a reform was clearly spelled out. And in all cases industry played in the historical development of the R&D system a particular not unproblematic role. But what are the possibilities and limitations of research institutions in the "steady state"? If there is something like a more global transition taking place then what direction does it take?

In principle two tendencies for change can be identified. The first is driven by influences external to the science system. Science has managed to gain such important place in our societies through technological developments and the expectations - although less euphoric than decades ago - in scientific contributions to the national wealth and welfare are still rather high. The idea that a well functioning science system would also help to develop a powerful economic system is - although doubts are risen - still very much present in the discussion. This notion of science as a wealth creating motor of course puts quite some pressure on the science system for efficiency and accountability. "Value for money" has become an important consideration when negotiating for funding.

Let us now come back to our initial question: What is the role of industry in the national science systems and in particular what role does it play in the reform discussion? It is clearly visible that along with science policies applied in other industrially developed countries the objective is set to link the academic sphere more closely or efficiently to the economy. This trend is often interpreted as a result of decreasing science budgets as said in the introduction, but it is also linked to the changing character of the economic environment which is increasingly technology oriented.

However when looking at the past developments in the CSFR and Hungary, we observe that the linking of science to economy was a central feature, but we also realize that this policy is not without any risk. In Hungary as well as in Czechoslovakia the need to have revenues and the dependence of shortage economy created rather negative tendencies for the science system. The need to integrate into the marked forced researchers to give up what they wished to do and to adapt their work to market needs. And the interest to cooperate turned out to be very one-sided. Academia needed industry for reasons of funding, but industry was not really interested in innovative and unconventional research. That means that there are numerous preconditions that have to be fulfilled in the scientific, economic and political systems of the respective nation states in order to make a strong linking of science to economy fruitful for all those who are involved. It means that the driving force, i.e. the demand for innovative research has to be carried by a dynamic, well structured and financially stable industry, while the research sector needs enough financial and creative free space in order to get organized. Basic science still should be a major component in this system. An additional difficulty in the Eastern European countries nowadays is that science and economy policies have to be reshaped simultaneously which created a dilemma of mutual interdependence that cannot easily be resolved. As the moment of dissolution of the existing structures and building up of the new ones is extremely sensitive to any interference it seems not without any risk to couple a reform in the science system to a radical change in the organization of industry - a view which was also expressed by K. Balázs who studied closely the economic situation in Hungary. And although the Austrian case is different in many respects it also turned out rather clearly that a precondition for a cross fertilization between the industrial and the academic sphere is only possible if both of them are able to work independently in a stable manner.

Science under these new conditions: Indeed there is no clear vision about the consequences of this ever stronger interconnection between research and industry neither among scientists themselves nor among historians or sociologists of science. Obviously it is extremely complex and there is still a lot of controversial discussion to come.

Notes

1. See for the general discussion Cozzens et al. 1990 and in particular Rip 1990
2. Ziman, 1990:23.
3. See for a discussion of this question Pavitt, 1991.
4. Etzkowitz, 1990.
5. For an overview on the development of the Hungarian science system see_for example: Darvas, 1989; Besseneyi, 1991; Balász, 1987; Szabo, 1990; Swain, 1991. For the discussion on the relation economy/science see Balász, 1987, 1988, 1989, 1991a u.b; Balász & Hare, 1989, Oakey et al., 1990. See also Batt, 1988.
6. Besseneyi, 1991:21
7. World Bank, 1990:49
8. Darvas, 1988: 169 ff.
9. Inzelt,1991
10. Schödl, 1988:48.
11. World Bank, 1991
12. Balázs, 1988:283
13. For an overview on the developments in the Czechoslovakian research system see: Dvorakova, 1990; Hrbek, 1991; Müller, 1988, 1990; Whightman and Rutland, 1991; Wolchik, 1991.
14. Müller, 1988.
15. For the Austrian research system see: yearly reports of the Minstry for Science and Research; Melchior 1990, 1993; OECD 1987; Langer, 1991.

References

Bessenyei, I. (1991), *Die Hochschulpolitik in Ungarn von 1945 bis zur Gegenwart, Erarbeitung der Problemstellung*, Manuskript

Balázs, K. (1987), "Snapshot '86. Changes in R & D in Hungary in the 1980s," *Magyar Tudomany 2.*

Balázs, K. (1988), "Marked-oriented scientific research and development after the economic reform," *Acta Oeconomica* 39 (3-4): 271 - 290.

Balázs, K. (1989), *The strategic tasks of scientific research, technical development and higher education. Research and Development Financing*, Manuskript

Balázs, K. (1991a), *Lessons from an economy with limited market function*, Manuskript

Balázs, K. (1991b), *"Transformation of the Science and Technology System of East European Countries in Transition from Centrally Planned to Market Economy: Hungary,"* Manuscript

Balázs, K. und P. Hare (1989), *The management of R & D in Hungary at the end of the 1980s*, Manuskript.

Batt, J. (1988), *Economics Reform and Political Change in Eastern Europe; A Comparison of Czechoslovak and Hungarian Experience*, Houndsmill - Basingstoke - Hampshire, MacMillan Press.

Cozzens, S. E., et al., Ed., (1990), *The Research System in Transition*, Dordrecht, Kluwer Academic Publishers.

Darvas, G. (1989), Hungary, G. Darvas, Ed., *Science and Technology in Eastern Europe*, Harlow, Longeman, 140 - 199.

Dvorakova, O. (1990), 'Struktur und Entwicklung des Wissenschaftspotentials der CSSR', W. Meske, Ed., *Wissenschaft der RGW-Länder, Berlin, Akademie der Wissenschaften der DDR*, Institut für Theorie, Geschichte und Organisation der Wissenschaft, 101 - 117.

Etzkowitz, H. (1990), 'The Second Academic Revolution: The Role of the Research University in Economic Development', Cozzens, S. E., et al., Ed., *The Research System in Transition*, Dordrecht, Kluwer Academic Publishers.

Hrbek, A. (1991), *Die Finanzierung der tschechoslowakischen Wissenschaft und Forschung*, Arbeitsunterlage.

Inzelt, A. (1991), *Science and technology indicators for policy making, Transition period in the field of information system*, Vienna - Bratislava Conference.

Kreuter, J. (1990), 'Probleme der Strukturpolitik in der CSSR', H. E. Gramatzki Klinger, F., Nutzinger, H.G., Ed., *Wissenschaft, Technik und Arbeit: Innovation in Ost und West,* Kassel, 197 - 203.

Láng, I. (1989), "Science and technology advice to government in Hungary," *Worldwide Science and Technology Advice* : 208 - 213.

Langer, J. (1991), *Austrian Universities in Transition - From Ministerial Control to Market Coordination*, Manuskript.

Melchior (1993), *Zur sozialen Pathogenese der österreichischen Hochschulreform*, Baden-Baden, Nomos Verlagsgesellschaft.

Melchior, J. (1990), "Zur österreichischen Forschungs- und Technologiepolitik: Entwicklungen und Probleme im Kontext internationaler Diskussionen," *Österreichische Zeitschrift für Politikwissenschaften 3*: 245-266.

Müller, K. (1988), 'Czechoslovakia', G. Darvas, Ed., *Science and Technology in Eastern Europe*, Harlow, Longman, 45 - 94.

Müller, K. (1990), *"Zur Perspektive der wissenschaftlich-technischen Zusammenarbeit zwischen Deutschland und der Tschechoslowakei,"* IGW-Report: 1

Oakey, R., et al. (1990), *"Diffusion of process innovations in an East European economy: some conceptual observations,"* Science and Public Policy 17(2): 97 - 104.

OECD (1987), *Review of National Science and Technology Policies*, Austria, Paris

Pavitt, K. (1991), *"What makes basic research economically useful?"* Research Policy 20: 109-119.

Rip, A. (1990), 'An Exercise in Foresight: The Research System in

Transition', Cozzens, S., et al., Ed., *The research System in Transition*, Dordrecht, KluwerAcademic Publishers,

Schödl, G. (1988), *"Zwischen Krisendruck und Reformerwartung, Zum Stand der Wissenschaftspolitik und Forschungsstrategie in Ungarn,"* IGW-Report 4/1988: 45 - 58.

Swain, N. (1991), 'Hungary', Ed., *Handbook of Reconstruction in Eastern Europe and the Soviet-Union*, Harlow, Longman, 59 - 145.

Szabo, M. (1990), *"Ungarns Wissenschafts- und Hochschulwesen im Prozeß der Demokratisierung,"* IGW-Report 3/1990, 4, Jg.: 61 - 71.

Wightman, G. und P. Rutland (1991), "Czechoslovakia," *Handbook of Reconstruction in Eastern Europe and the Soviet Union* : 29 - 58.

Wolchik, S. L. (1991), *Czechoslovakia in transition*, London, Pinter Publishers.

WorldBank (1990), Report on Hungary.

Ziman, J. (1990), 'What is happening to science?', Cozzens, S. E., et al., Ed., *The Research System in Transition*, Dordrecht, Kluwer Academic Publishers.

11 Opinions and attitudes on technology, science and labour in reunified Germany

Dieter Jaufmann

"The relationship between technology and society is a crucial matter for contemporary social thought and practice. Whether it is the socialist East or the capitalist West, the development, acceptance, and use of technology are seen as an issue of paramount importance for the progress of society" (Mills 1986: VII)

Aims, objectives , problems

After the 'Wende' in November 1989 and the reunification in October 1990, Germany has become a gigantic elaborateatory. Both parts of Germany and the people living within have been divided for more than 40 years, with different political, economic, and social systems. All these years a border between East and West parted Germany, and, symbolically the world.

From the beginning of this field-experiment, now the biggest in the world (see for example Giesen/Leggewie 1991), western style opinion polling has spread East. The greed for, and the power of figures resulted in a gold-rush-atmosphere (see Jaufmann/Kistler 1991: 108 ff.). A variety of problems came about by polling this way. To give an impression of the difficulties I will name a few of them:

- People in the former GDR had really very little experience with polls, especially with representative ones. They were not used to this instrument. The only institution which has done such special kind of work - the 'Institut fur Meinungsforschung' in East Berlin - had been closed by politicians at the end of the seventies.

- The recruitment of the 'right' field-staff was problematic.

- The infrastructure for communication (post, telephone, etc.) was insufficient. It took a long time to conduct the field-work.

- Sampling was difficult. There was no ADM-mastersample and also no basis for random-sampling.

- Collecting data by telephone-interviews - which had been done very often in the early 'wild' times - was pure nonsense; only 15-20 % of the people there had a telephone. Furthermore, it should be asked who the ones were who had a telephone, and especially, why just them?

- Using the same questionnaires, without checking out if the understanding and meaning of the wording is the same as in the West, i.e. neglecting semantic problems.

All in all I therefore agree with the following statement from Karin Sieger, a pollster working at the MORI-institute in Great Britain:

> "While one cannot help but acknowledge that there may be examples where economic profitability might have been favoured over the adherence to professional standards, it is highly inappropriate to discredit the polling industry per se, without looking at different research techniques employed and making reference to the special circumstances Western researchers are faced with when polling in East Germany" (1990: 324).

Despite these problems and limitations it makes sense to know what people think, which opinions and attitudes they have. This seems to be especially true if one thinks about the so called 'Thomas-theorem': If people define situations as real, they are also real in their consequences and the basis for their acting.

The selected results that will be presented and discussed in the following article have been generated by a special kind of secondary analysis (c.f. Jaufmann/Kistler/Jdnsch 1989: 10 ff.). That means in one case focusing on a lot of polls, examining the methods that have been used critically, and then comparing the results. The consequence is a considerable and secure range of information, a valid 'tunnel' that is based on more than just one 'spot-light-poll'. All figures and results presented here are, of course, only examples. They are however typical, and stand for many others.

The paper deals firstly with attitudes towards technology on the general level in the East and the West of Germany. In the following chapter how the people see the role of the state concerning science and research will be discussed. Next, I will focus on the attitudes towards labour. The paper will be completed by discussing some problems of the so called 'Transformation

Research' and a few concluding remarks and hypotheses.

Attitudes towards technology after the 'Wende' in the eastern and the western federal states of Germany

Similar to the USA, Canada, Great Britain, Japan and Australia, the people in the former GDR had a very close link between their attitudes towards science, and their attitudes towards technology. Both terms were often used together: 'scientific-technological progress' or 'scientific-technological revolution' are examples. In the Western parts the two concepts vary from each other.

As we have demonstrated in another study, people have an abstract and overall balancing attitude towards technology (see Jaufmann/Kistler/Jdnsch 1989). As of now, real time-series of the indicators in an East-West comparison cannot be found. Only a few replications of some of these questions have been made. Therefore special weight has to be put on the variety of indicators, the figures of implemented polling-institutes, and a broad secondary analysis.

Meanwhile, classical (since 1966) is the question of the 'Institut fur Demoskopie Allensbach', if technology is more a blessing or more a curse all in all. In Autumn 1990 this question was used in an Emnid-poll. The results were almost identical in the old and the new Federal States. In both cases the positive item tended against the 60% margin. Furthermore, in both parts, one out of ten has seen technology more as a curse. The rest decided for the item 'Don't know'. The results fit together nearly perfectly. The same result came out from the traditional overall balancing indicator of Emnid in this field. The wording of this question was the following: "Do you believe that technological progress, for example computers, screens, microprocessors taken all together, will help or harm mankind?". Cumulated over three waves of interviews (from the end of November 1990 until January 1991) around three quarters of the people in the East and the West, believe that it will help. In this case also, every tenth person decided for the opposite. Because of the only bipolar opportunity for answering (help/harm), the judgement was still more positive than in the case before. The part of the sceptical-opposing people remained identical.

One further question on the general attitude-level towards technology, was used by two pollsters. Question-wording was exactly the same and therefore of special interest for our research. This also seems to be true and especially from a methodical point of view. The exact wording was the following: "How would you classify your opinion towards technology on the general level. Please make your judgement based on this list. No. 1 means: 'Your personal attitude is very negative', No. 7 means 'Your personal attitude is very positive'. With the figures in between, you may up grade your opinion".

The results for this question are shown in figure No. 1, which show both survey-institutes and also the East-West-split in every case. The similarities are really astonishing. Emnid traces around 10% sceptical-opposing, 20% ambivalent and 70% with positive attitudes towards technology.

Compared with the Basis Research results, from eight months later, the data shows that the differences are rather marginal: The positive thinking part of the people on this topic now seems to have increased a little. The same question was asked by Basis Research one year later in Autumn 1992. In the East and the West the results remained nearly the same as in 1991, i.e. the percentages for each of the seven numbers were still identical.

Of course, not all indicators from all polling-institutes measure the same way and show identical results. Before I summarize all the data of our secondary analysis and draw some general conclusions, I will give an example where differences arise. The IPOS-institute asked people in which society they would like to live. For our purposes the relevant bipolar pair of answers read following: 'In a society which looks more sceptical towards technological progress, because it will bring great danger in the future' (= figure No. 1 on the scale) or 'In a society which promotes technological progress, because it will secure the future' (= figure No. 7 on the scale) (see IPOS 1992: 19 ff.). In the years 1984 to 1992 the means for the old FRG were highly constant between 4.5 and 4.2; in the years 1990 to 1992 it was 4.4. The comparable results for the new Federal States were 5.8, 5.5 and 5.4; that means in each of the three years they were (more than) a whole grade higher. People there hold the stronger opinion, that society should promote technological progress. Maybe this depends on a real lack of wide spread modern technology, and they have to catch up now. Maybe technology is also seen as the key issue for the modernization of society.

Taking all the results, not only the ones presented here, into consideration, one can conclude that the attitudes towards technology on the global-balancing level in the East and the West of Germany are very similar. This result is valid to the greatest possible extent, as shown by all kinds of questions, surveys, and polling-institutes. Also in the differentiation by gender, the results are quite the same. The attitudes of men are more positive, and women are more sceptical. Furthermore, women are also ambivalent to a greater extent. These results can also be found in international comparisons, now and going back in time (see Kistler/Jaufmann 1989). Seen under the perspective of the highest reached formal education, however, differences appear in most cases. With a rising grade of education in the new Federal States, we also find rising positive attitudes towards technology. In the Western parts, however, the course is unhomogenous, even contrary in parts.

How people see the role of the state towards science and technology

Not only the people's own attitudes towards technology and science are of interest and important for a modern society, but also what they think the state and the government should do in this field. Should the state intervene, and when, and in which cases? This was one of the questions that the Infratest-institute asked at the end of 1990 (for the detailed results see table No. 1). In the eyes of those people asked in the East and West, this is very similar. About one out of ten people fundamentally opposed the fact that the state should intervene in scientific and technological research. They believe it will be self-regulating if necessary. Nearly 40 % also believe that the state should not directly intervene in this sector. At the same time however, concerning the use of the invention they plead for a regulation set by democratic law. 42 % in the West and 47 % in the East declared the state should forbid research that could be dangerous from its results to mankind.

A similar question has been used by Emnid (see table No. 2). There are only two possibilities to decide: 'total freedom for science and technology' versus 'careless control is necessary, because of possible dangerous results to people'. Within the borders of the former FRG, one third decided for the opportunity of 'freedom', while in the East 27%. Concerning careless control, 66% in the West voted for it and in the East 72%. These results also show, that there is more preference for interventions of the state/government in the East. The same is true for women in both parts of Germany.

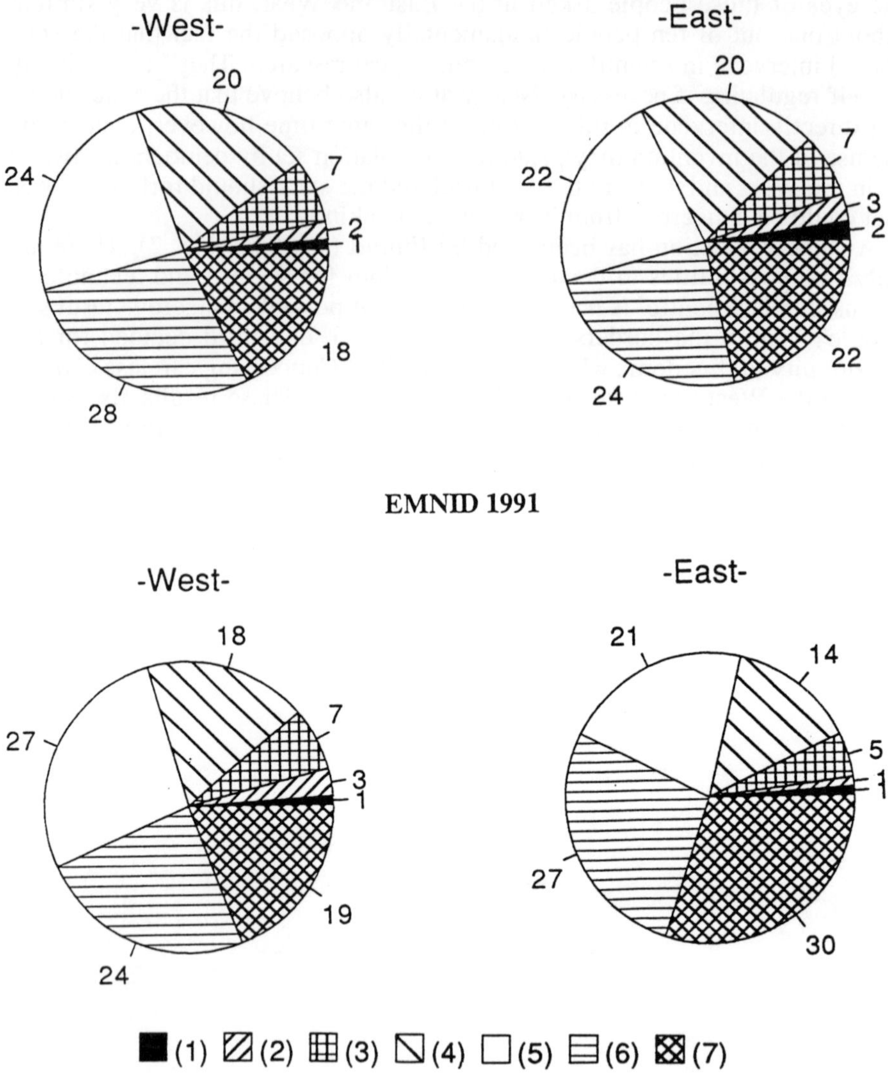

Legend: (1) = very negative, (7) = very positive

Figure 11.1 Personal attitudes towards technology on the general level (in percentage)

Table 11.1
Technological and scientific research: How should the state act or react? (figure given in percentage)

	State should not intervene. Science and research are free and self regulating	State should not intervene. If necessary, they should, by democratically set law, forbid the use of intervention.	State should forbid research whose results could pose a danger to mankind	Don't know	No answer
Total	11 (11)	41 (37)	42 (47)	6 (4)	0 (1)
Sex					
men	9 (13)	46 (40)	42 (46)	3 (1)	- (0)
Women	13 (8)	36 (35)	42 (48)	8 (6)	0 (2)
Age					
14-19	10 (12)	41 (26)	47 (53)	1 (4)	- (6)
20-29	8 (9)	41 (40)	48 (49)	3 (1)	- (0)
30-39	8 (11)	38 (38)	51 (49)	3 (2)	0 (-)
40-49	10 (6)	51 (42)	37 (51)	2 (2)	- (-)
50-59	9 (12)	46 (34)	40 (47)	6 (6)	- (0)
60+	17 (13)	34 (38)	36 (38)	13 (8)	1 (3)
Education					
Primary	14 (14)	33 (34)	45 (43)	9 (7)	0 (2)
Secondary	8 (8)	44 (38)	45 (52)	4 (1)	- (1)
Tertiary	9 (7)	57 (53)	32 (39)	1 (1)	0 (-)

Legend: X = West; (Y) 0 East
Source: authors own compilation from: Infratest (1990), Deutschland 2000, Munich: mimeo.

Table 11.2
Science and technology: 'Total freedom' or 'careless control'?
(percentage figures)

	Total freedom for science and technology	Careless control of science and technology necessary, because of possible dangerous causes	No answer
Total	33 (27)	66 (72)	1 (1)
Sex			
men	35 (31)	64 (69)	1 (0)
women	31 (24)	68 (74)	1 (2)
Age			
14-19	20	78	1
	}30(27)	}67 (71)	}3 (2)
20-29	35	62	3
30-49	33 (26)	67 (73)	1 (1)
50-64	37 (23)	63 (77)	0 (-)
65+	32 (32)	67 (65)	1 (2)
Education			
Primary	34 (30)	65 (67)	2 (2)
Secondary	33 (25)	67 (75)	0 (1)
Tertiary	30 (28)	70 (71)	0 (0)

Legend: X = West; (Y) = East
Source: Authors own compilation from: EMNID (1991), Umfrage & Analyse 1, Bielefeld.

These two examples illustrate a tendency, which is valid for almost the whole topic 'science, technology, and research' that people from the new Federal States believe more in interventions from the state than in the Western part. A consequence from more than 40 years of centralistic orientation, education, and thinking?

Attitudes towards labour

To summarize, the results of every survey known to us and conducted after the 'Wende' dealing with the topic 'labour, jobs, etc.' have shown an important, and in most cases even an overwhelming place of this value for the people in the East. But also in the West it is given a high rank in priority-lists. The reason seems to be on one hand a high social moral of labour, on the other the capitalistic ethics of work. This seems to be true in history and also right now. Furthermore, we have to consider, that in the former GDR there was not only a duty to work but also a right to work. That was even written in Article 24 of their Constitution. The employment-rate for women was one of the highest in the world: More than 90% of them were employed in paid jobs at the end of the '80s (see Nickel 1990:10). Unemployment was a real new and shocking experience for the people in the Eastern part of Germany after the 'Wende'.

There have been and still are a lot of questions concerning the various aspects and facets of labour. In the following chapter, I will concentrate mainly on the question, how important is work for the people. The results in table No. 3 have been gathered by three surveys in the years 1990 and 1991 (see Hausstein/Gvtze 1991). In two cases they represent the Eastern part of Germany in total and in 1991 there was also a survey only for the Federal State of Brandenburg. It becomes obvious, that in all three polls more than 90% stated that labour was either important, or even very important in their life. The value for 'very important' was still rising from 1990 to 1991. There are also little differences in the results of the five new Federal States and the Federal State of Brandenburg surrounding Berlin.

The Socio-Economic Panel (SOEP) meanwhile has been gathered since 1990 in the Eastern part. When asked how important work is for the subjective well-being in 1990, 53% of the people decided for the item 'very important' on a 4-point-scale (see Landua, 1992:12 ff.) The repetition of the question one year later resulted in 58%, this means that work has become more important. The tendency here is also the same than shown in the surveys above.

Table 11.3
Individual importance of labour (in percentage)

		Federal State of Brandenburg	
	1990	1991	1992
very important	67.7	81.1	83.1
important	25.0	14.7	13.9
fairly important	2.0	0.9	1.0
less important	1.2	0.7	0.5
unimportant	1.2	1.4	1.0
no answer	2.9	1.2	0.6

Note: The surveys were only conducted in the five new Federal States or in the Federal State of Brandenburg. Figures here are unweighted, but differences between weighted and unweighted figures in 1991 are only very slight.
Source: Authors' own compilation from: Hausstein, B./H. Götze (1991), Leben '91. Daten und Feldbericht, Berlin: mimeo.

Table 11.4
Evaluation of labour and leisure-time (in percentage)

	Job more important than leisure time			equal			leisure time more important than job		
	West		East	West		East	West		East
	1984	1988	1990	1984	1988	1990	1984	1988	1990
Total	30	27	39	40	44	39	30	29	22
Sex									
men	33	27	42	42	47	39	25	26	19
women	24	26	36	36	38	39	40	36	25
Age									
18-24	19	17	32	38	42	27	43	41	41
25-34	23	20	39	41	45	38	36	35	22
35-49	35	31	42	39	45	41	26	24	17
50-65	37	33	39	41	43	46	22	25	15

Source: Habich, R./H.-H. Noll (1992), Einstellungen zur Arbeit und Arbeitszufriedenheit, in Bundeszentrale für politische Bildung (Eds.), Datenreport 1992, Bonn, p. 584.

Labour in the former 'workers society of the GDR' of today compared with the Western Federal States of Germany, is seen by the people as more important and of higher value. This can also be demonstrated by the results from the baseline-study of the ALLBUS in 1991 (see ZA/ZUMA 1992). 'Profession and labour' is regarded as 'very important' in the West by 39% compared to 64% in the East (measured on a 7-point-scale). The proportion of all the three positive figures summed up was 84 to 90%.

Our last exemplary data-base to be presented here is the 'Welfare-Surveys'. The question 'what is more important: work- or leisure-time?' seems to be of special interest to us. For the territory of the old Germany we have the results from two polls from 1984 and 1988. As you see from table No. 4 there has not been much change in the figures in between. On the total level around 30% found their job more important than leisure-time. On the other hand also 30% found their leisure-time to be more important than their job. For the remaining 40% both, job and leisure-time are of the same importance. More men than women valued labour higher than leisure-time. On the other hand, women showed higher preference for leisure-time. The classification by different age-groups shows a rising preference for jobs with increase of age. The same group-specific tendencies are to be found in the eastern results from 1990. The main difference is, that people in the East are clearly more strongly oriented towards their jobs, and less towards free time. The only real exception is to be found for the age-group of 18-24 years: They have a very strong preference for a job, but the same values of leisure compared with the corresponding age-group in the West.

Facts and figures - some problems of so called 'transformation research'

As I have shown in the three chapters about technology, science and the role of the state, and also labour, the attitudes of the people in the East and the West of Germany are amazingly similar in a lot of cases. However there are also considerable differences. A fundamental one in the group-specific differentiation and view, is, for example, the increase of positive attitudes of the people with their rising grade of education in the new Federal States. In the West this connection is unhomogenous and it is different for diverse kinds of technology and sciences.

For a better understanding of these similarities and differences in the attitudes of the population that have developed from living in different political, economic, and social systems for more than forty years, further research is needed. The following points are particularly important:

- to look back at the time before the reunification,
- to secure the old empirical data of the former GDR,
- to secure their interpretation methods,
- to be sure of the method which generates the empirical material, that means critical examination.

All of which is easier said than done. I will illustrate this with a few remarks. Going back in history shows that the social sciences in the GDR in general, or in most cases, have been stripped of their intellectual freedom and potential. There have been no public social sciences, but they were a domain of the SED-party right from the start. Sociology for example was a controlled science and most of its results were 'under control' and secret (c.f. Jaufmann 1992: 76 ff.). Since the 'Wende' a lot of the empirical data of social research has been thrown away or has been 'privatized'. The evaluations of the institutes of the Academy of the Sciences of the GDR, also of the universities, and afterwards often their suppression foster this process. We have the same problem with the interpretation of empirical data. Figures don't speak for themselves; we need to know what is behind them. Why, how, and on behalf of whom have they been collected? In a lot of cases, the people who have gathered the data and therefore have the necessary 'background-knowledge' now have changed their occupation. Some moved from East to West and others have not yet been asked nor ever been mentioned. These facts inevitably result in a loss of knowledge and the information from the data is sometimes vague. We have the same kind of problems with the methods practised so far. In this sense 'Transformation Research' comes near to the work of detectives.

Concluding remarks and hypotheses

The end of the GDR was also the end of the old FRG. The processes now are no one-way transfer from West to East. The presently developing Germany will be based on a variety of traditions and will go on to an open future. This open society will not be the result of a formal addition of components, neither will it be a co-evolutionary distribution of power and dominance.

This future is to be seen in the context of the rapidly changing societies in the other Eastern European States. Also to be considered must be the integration of the European Community and its common market coming into force in 1993.

In the centre of this transformation and modernization process still stands the individual. In order to take this into account, not only economic developments and processes are necessary, but also mental ones. Both these will take a long time. In the years since the reunification in 1990, people in the East and the West of Germany have become more and more pessimistic about the question, how long it will take to have the same standards of living as in the former Western part of Germany. Common belief (1992/93) is that this process will take ten years, or even longer (c.f. o. V. 1993: 52 ff.; Forschungsgruppe Wahlen 1993: 1). This means for example that in the view of the Eastern people, the estimation of the time needed for this purpose has almost doubled in two years. Furthermore, it seems to be that the people in the East and West are growing more and more apart (c.f. e.g. o.V. 1993: 52 ff.). If this alienation and division continues in the future, the result will be a new wall between the East and the West Germans - this time a mental one in the heads of the people.

Having considered all available evidence, one thing remains certain: The flourishing gardens and landscapes which were to have come into bloom, as predicted by Chancellor Kohl and other politicians, will need more time than first expected.

References

Forschungsgruppe Wahlen (1993), Ostdeutschland, 'Der neue Realitätssinn', in *INTER/ESSE 1*, p. 1.

Giesen, B./C. Leggewie (1991) (Eds.), *Experiment Vereinigung. Ein sozialer*

Großversuch, Berlin: Rotbuch.
Hausstein, B./H. Gvtze (1991), *Leben '91. Daten und Feldbericht*, Berlin: mimeo.
Inglehart, R. (1989), *Kultureller Umbruch*, Frankfurt/Main, New York: Campus.
IPOS (1992), *Einstellungen zu aktuellen Fragen der Innenpolitik in Deutschland*, Mannheim: mimeo.
Jaufmann, D. (1990), 'Technik und Wertewandel - Jugendliche und Erwachsene im Widerstreit?', Frankfurt/Main u.a.O, Lang. Jaufmann, D. (1992), 'Empirische Sozialforschung, Umfragen und Soziologie in 40 Jahren DDR - Die 'Wuste' lebte!', in H. A. Henkel/U. Merle (Eds.), "Magdeburger Erkldrung" - *Neue Aufgaben der Wohnungswirtschaft. Beiträge zum deutschen Einigungsprozeß*, Regensburg: Transfer, p. 73 ff.
Jaufmann, D./E. Kistler/G. Jdnsch (1989), *Jugend und Technik. Wandel der Einstellungen im internationalen Vergleich*, Frankfurt/Main, New York: Campus.
Jaufmann, D./E. Kistler (1991), 'Bestandsaufnahme und Perspektiven der Einstellungs-, Markt- und Meinungsforschung in einem vereinten Deutschland. Bericht über eine Tagung', in *ZA-Information 28*, p. 108 ff.
Jaufmann, D. et al. (1992) (Eds.), *Empirische Sozialforschung im vereinten Deutschland. Bestandaufnahme und Perspektiven*, Frankfurt/Main, New York: Campus.
Kistler, E./D. Jaufmann (1989), *Geschlechtsspezifische Unterschiede in der Technikakzeptanz - Ein Survey vorliegender Umfrageergebnisse*. Report for the Federal Ministry for Economy, Stadtbergen/Leitershofen (December): mimeo.
Kistler, E./K.-H. Strech (1992), 'Die Sonne der Arbeit - Arbeitseinstellungen als Forschungsgegenstand im Transformationsprozeß', in D. Jaufmann et al. (Eds.), *Empirische Sozialforschung im vereinten Deutschland. Bestandsaufnahme und Perspektiven*, Frankfurt/Main, New York: Campus, p. 155 ff.
Landua, D. (1992), *Das Jahr danach*, WZB-Paper P 92-102, Berlin.
Maaz, H.-J. (1990), *Der Gefühlsstau. Ein Psychogramm der DDR*, Berlin: Argon.
Mills, St.C. (1986), Preface, in F. Adler et al. (Eds.) *Automation and Industrial Workers. A Cross-National Comparison of Fifteen Countries*, Vol. 2, Part 1, Oxford: Pergamon Press, p. VII ff.
Nickel, H.-M. (1990), '... kommt an den Herd', in *Sonntag 8*, p. 10.
o.V. (1993), 'Erst vereint, nun entzweit' SPIEGEL-Umfrage über die Einstellung der West- und Ostdeutschen zueinander, in DER SPIEGEL 3, p. 52 ff.
Sieger, K. (1990), 'Opinion Research in East Germany. A Challenge to Professional Standards', in *International Journal of Public Opinion Research 4*, p. 323 ff.
Zentralarchiv für empirische Sozialforschung (ZA)/Zentrum für Umfragen, Methoden und Analysen (ZUMA) (1992), ALLBUS Basisumfrage 1991 in

Gesamtdeutschland. Codebuch, Köln, Mannheim: mimeo.

12 European science and technological development

Peter Mettler

Introduction

Although the EEC offers the services of a central data bank[1] for information on European Science and Technological Development (STD or ESTD), I venture the thesis that there is nobody who could come up with a comprehensive analysis of ESTD. A couple of explanations can be offered:

1 The numbers involved are simply too great.[2]

2 ESTD consists not only of what is financed or supported by Brussels (the EEC), but includes the wealth of all the other state and private STD co-operations, be it between university researchers, state owned research labs or between industry (bi-, tri- or multi-lateral). Institutions such as the following merit particular attention: the Council of Europe, the European Science Foundation, the European Cultural Foundation and the European Industrial Research Management Association. And there is also the wealth of research on ESTD undertaken by third institutions like UN´s ECE-Economic Commission on Europe, OECD, here in particular the many "Reviews of National Science and Technology Policy", NATO, etc..

ESTD could be defined in two further ways:

1 STD should only be called ESTD, which contributes something to Europe.

2 ESTD will only then acquire the notion of being something specific, if it can clearly be distinguished from developments e.g. in the USA or Japan.

Consequently, this paper can only have a narrow goal: to introduce a small number of research and/or technical development programmes, to ask, if they should merit being called European, and to raise the question whether or not by implementing ESTD, an European identity could be fostered.

1. EEC (i)

In the first stages of the EEC, STD concentrated on coal and iron. Gradually it extended into the field of energy in general. In 1972, Ralf Dahrendorf provoked the first discussions on Europe´s future, in particular on STD, which resulted in the "Europe plus 30"-project. Its report - also known as Kennet-Report - was delivered in 1975. The FAST-project[3] began in 1978 and is now part of the MONITOR programme (1989-1992). In 1983, the first EEC Framework-Programme for STD was approved by the Council of Ministers. Its most renown sub-programme was ESPRIT, now in its second phase (1987-1993), whilst its symbol is still the JOINT RESEARCH CENTRE (JRC), which unites today eight research centres[4] in four countries with some 3000 researchers and a budget of 4.5 bill. ECU. The first Framework-Programme had a volume of 3.75 bill. ECU, the second a volume of 5.396 bill. ECU and the third, which will end in 1994 or earlier, a volume of 5.7 bill. ECU. This really is a small sum, just 6% of the overall amount spent in all member countries on STD. The German Ministry for Science and Technology (BMFT) already has some 4.5 bill at hand. ECU. Preparations for the fourth EEC Framework-Programme for STD (1993-1997) are in full swing, MONITOR/FAST will contribute to it considerably (see below).

The following considerations owes a lot to Matthias Tang´s article "Die Forschungs- und Technologiepolitik der Europäischen Gemeinschaft" (S&T policy of the EEC), in: "wechselwirkung", June 1992:

a ESPRIT-European Strategic Programme for Research and Development in Information Technology was developed jointly by the Commission and the 12 most important European companies in electronics. The annual plans are developed jointly by the Commission, the ESPRIT Advisory Board and the ESPRIT Management Committee. The main tasks are: the development and application of norms, software and of the highest integrated chips. A precondition for every project is that at least two companies from two different countries collaborate and furnish 50% of the research budget. Two different types of projects exist:

a(i) Projects for the development of long term strategic technologies.
a(ii) Short term, small scale projects.

Seventy five percent of ESPRIT II´s 1.6 bill. ECU budget go into type a, i.e. serve large industries who are able to wait for the return on the investment. The rest goes into type b and to small and medium sized industries with fast return necessities.

b With reference to decision procedures concerning ESTD, decisions concerning:

b(i) framework-programmes, have to be distinguished from decisions concerning
b(ii) research projects.

In b(i), the Council of Ministers for STD comes to a decision after the Commission has submitted proposals. Although the European Parliament also discusses the Commission´s proposals, it can hardly alter it nor does it have finances available.

In b(ii) the procedure is inductive and hardly controllable. Numerous committees (with numerous methods of how to appoint members), above all the IRDAC (Industrial Research and Development Advisory Committee) form Brussels´ lobby. Interest-groups, unable to maintain a permanent representation in Brussels, can hardly influence any decision. The fact that the two Germans within IRDAC are the heads of RTD at Siemens and at BASF could be taken as an example for large industry´s lobby power in Brussels.

c Numerous examples demonstrate that, as soon as one tries to adhere just slightly to the three official guiding principles of all EEC-STD-policy, piles of difficulties arise. These three principles are:

1 To assure competitiveness.

2 To act only as a subsidiary, if national solutions no longer make sense.

3 To give priority to basic research.

One of the most famous cases in question is the conflict between assuring competitiveness in comparison with e.g. Japan and stimulating competition within the EEC (or to stimulate an even distribution of industry throughout the EEC).

A second one can be described as the dilemma between subsidiarity and solidarity: If only five member states dispose of e.g. research facilities for one specific technology, and if these five realise that they would need an EEC subsidiary in order to assure their competitiveness, should the EEC then establish such a subsidiary or much rather supply comparable research facilities to all those member states, who do not yet have such facilities readily available?

And finally, a third question. What does basic research really mean today? Does basic research today not require an infrastructure only available in the (industrial) centres? And how is it possible to develop an industrial infrastructure in the European periphery (e.g. in Portugal), if the centres of STD have to stay where they are?

All honest examinations of ESTD came to the conclusion that, despite the big European Multinational Corporations (EMC) being the net winners of ESTD, the EEC has to and does devote considerable funds to inner community cohesion. But the EMCs are not satisfied, particularly not in most areas of high-tech. One of the remedies developed was EUREKA-European REsearch Koordination Agency. It started in 1985 under the premise of being the direct European answer to the funds spent by the US for the SDI programme. EUREKA is All-European in as far as it is open to EFTA members as well as to Turkey (besides the twelve EEC member states and the European Commission), but EUREKA is just one more organisation outside the EEC (as are e.g. the European Space Agency and the Airbus Consortium), which serve as levers to the big European states (and their EMCs) to steer the inner-EEC negotiations. And no mention has been made so far of military research and technology and/or armament industry.

2 The European Science Foundation

Opposite to EEC activities, the European Science Foundation (ESF) now counts 59 research councils from 21 countries as members. It disposes of some 2.5 bill. ECU, which is quite considerable for an organisation which is not running its own research laboratories. In order to describe ESF´s mission, some quotes from ESF´s vice-president William Mitchell might enlighten the situation:[5] "Three aspects are relevant to a European approach:

(i) when the project is very expensive and costs have to be shared;

(ii) when the subject is wide or covers many sub disciplines which an individual country is not likely to cover completely;

(iii) when the subject itself deals with 'Europe' ".

ESF is not directly involved in the first category. The second is central. It is an interdisciplinary role and great strength is added by introducing the sub disciplines to each other across Europe. Finally a European approach is relevant when the subject involves Europe, as with the Geotra-verse, Environment, Science and Society, Language Typology, etc. Clearly the challenges facing the ever widening Europe, with the dissolving of barriers between west and east and the break-up of the former Soviet Union, pose problems for science and research. We must be alert to these problems and make sure that the people served by the organisations the ESF represents make their voices heard. The history of the ESF makes me confident that it is a body whose authority will help to transmit their voices, whether they relate to the level of scientific support or the disposal of nuclear material.

Increasingly, governments regard the ESF as a Europe-wide source of independent and objective advice. The advice offered by ESF to the

Ministerial Meeting of the European Space Agency was well received. Similarly two of the ESF Committees are run jointly with the EEC (Environmental Change, and Ocean and Polar Research), which has also commissioned a report from ESF on social science research.

Nobody should be surprised that there are many links between the EEC (and the Commission in particular) and the ESF. One of the links of great interest here is the mentioned report "Social Sciences in the context of the European Communities", written by H.Newby, H.Nowotny, A.Alsop and J.Smith and published in 1991. The authors were asked to "consider the background, definition and illustration of possible social science contribution to Europe" and found, that European social science research almost entirely lacks Community awareness. And they concluded that it might be the greatest contribution of European social science research to propose ways and means, how to increase this awareness as well as possible European Identities.

Browsing through some of the general leaflets of ESF, describing its activities, the following is found to be repeated manifold:

> "- ESF works for the advancement of European science;
>
> ... ESF´s experience in developing cost effective means of scientific management on a transnational scale, ESF has more ´connecting power´ over a wider range of countries and basic disciplines than most other agencies".

The best example for this activity and intention is the programme "European Research Conferences", started in 1990 and largely modelled on the Gordon Research Conferences in the USA. But besides the obvious advantage of a co-operation between bodies which are geographically close to each other, the question remains whether or not the term "European" has any further meaning? A question which seems even more worthwhile asking, since these conferences are open to scientists from all over the world.

Umberto Colombo´s "President´s Report" in ESF´s "Annual Report 1991" writes that"it is right that ESF should seek a role which ... assists European science as a whole in its contacts with leading scientific bodies world-wide. ... Collaboration with leading foreign institutions not only adds lustre to the Foundation, it also reflects well on the standing and reputation of European science".

"A key issue is the extent to which in Europe it remains optimal to finance nationally most of basic science. There is an increasing tendency for research funders to look for wider areas of co-operation or concentration. There are several factors at work, but a prime mover is globalisation of scientific knowledge itself. ESF Member Organisations are increasingly operating on a level which transcends national frontiers, especially when added value is to be gained from co-operation and synergy" (p.2).

One of the answers could be what can be found in the report on "The

ASCEND 21 Science Conference"[6] in: ESF-communications, no. 26/1992, p.23:

"Modern society has become an intricate mass of highly complex systems and, in the effort to dominate them, we must ever more rely on capabilities deriving from the whole spectrum of our knowledge. Global problems ... demand the harnessing of all of human knowledge. Interdisciplinarity is the only way by which man can first comprehend their significance, and then seek solutions. ... A social conscience must evolve, and this brings us into the sphere of ethics. ... to ´leap-frog´ directly to solutions involving technologies which are currently available only in the most advanced countries, and which can be grafted on to traditional technologies, leading to appropriate technological blends. ... In the conventional idea of the industrial society, increasing consumption of manufactured goods is in practice a prerequisite for sound growth of the industrial sector and is regarded almost as if it were a moral imperative. Communicating the message that development must be sustainable and therefore sparing, efficient and frugal, is no simple matter. ... we have to set an example, but this implies a radical change in our own culture and lifestyle, and a new spirit of solidarity as citizens of the world. ... Moreover, the essentially international dimension of modern science ... may help a new international approach to global problems to evolve".

3. EEC (ii)

Drive

One of the most well known and at the same time most voluminous EEC programmes is the DRIVE-Dedicated Road Infra-structure for Vehicle Safety in Europe-programme, now in operation for almost five years. Its intention is to come up with a European IRTE-Integrated Road Traffic Environment on the basis of a system of RTT-Road Transport Telematics (in the first years it was still RTI-Road Transport Informatics) and it seeks to improve road safety, to maximise road transport efficiency and to improve the environment.

In this year´s operation (Operation 1992), some 2000 individuals and organisations participated. Though the general principle for DRIVE projects is the same as for most other EEC projects: costs are shared between the EEC and private companies or research institutions on a 50:50 basis, DRIVE is a particular case because of the intended system scale economics as well as the time perspective involved:

"(At the beginning will be) the conceptual design of a structure involving communication links and nodes, both fixed and mobile with an architectural complexity. Implementation is likely to be staged:

starting with one service (may be an up-to-date traffic control system) covering a limited area; continuing with the expansion of the area or the linear coverage of a corridor; adding additional network functions (route, guidance, weather alerts, parking and tourism assistance, co-operative driving aids, and so on); increasing the number of equipped vehicles; and increasing the capability of the in-vehicle equipment to exchange information with the roadside and other vehicles, and to process and present information to the driver. The system must be planned and implemented so as to minimise premature or negative investment (both fixed and mobile) and to permit staged expansion of capacity to match both the vehicle traffic and the communication traffic ...

The aim is to meet the various objectives of European society in responding to traffic congestion, improving efficiency and safety of movement, and limited adverse environmental impacts including the waste of natural resources" (Source: Operation 1992: Road Transport Telematics - Consolidated Work - Volume I - Page 3, Ref.: XIII / F5/DR 1480/1).

Fast

Dahrendorf´s ambition had already been, historically speaking tough from a modern point of view, to contribute to Europe´s future identity. But the more FAST developed, the more pragmatism took over and matter-of-fact studies prevailed over futures research:

"The aim of the FAST programme remains the multidimensional analysis of scientific and technological change so as to highlight their implications and consequences for the common R+D policy and for other Community policies. The new programme should therefore propose timely priority options for new long-term orientations for Community action in the fields of research chosen by the Council of Ministers, namely:

(i) relation between technology, employment and work,

(ii) transformation of services and technological change,

(iii) the emergence of new strategic "industrial" systems in the fields of communication and of food,

(iv) integrated development of renewable natural resources.

FAST´s other major task during the coming four years is to strengthen the foundation of prospective thinking in Europe by multiplying European co-operative networks between centres of research in

Community countries and national institutions responsible for long-time work" (Source: FAST Objectives and Work Programme 1984-1987, Commission of the EC, Brussels, February 1984, p.4, XII-201-84-EN).

It would be too long to follow FAST´s history in detail in a paper like this. The most decisive developments since the cited document was: 1. that FAST´s activities provoked the creation of two further institutions of futures research within the EEC: the "Prospective Cellule" close to the Commission´s Presidency, and the Institute of Science and Technological Prospective (PROMPT) within the Common Research Centre at Ispra, and 2. that FAST was integrated into the wider perspective of the MONITOR programme (1989-92), which included strategic and evaluative analyses (SAST-Strategic Analysis in Science and Technology and SPEAR-Support of the Evaluation of R&D Programmes). The Official Journal (C 144/5, dated June 6, 1989) formulated:

> "The FAST activity is a reoriented follow-up of previous FAST programmes. It will study the scientific and technological developments and the interactions with economic and social changes in the Community in the light of world-wide developments. The aim is to provide the Community with global long-term analysis in relation to the Community´s major objectives in the 1990´s, namely the creation of a single internal market and the strengthening of economic and social cohesion".

The two following citations show, however, that this reorientation was not meant to be abrupt. But FAST's pronouncement stating that the EUROPROSPECTIVE III-conference with the title "Global Perspective 2010, Tasks for Science and Technology" will take place in June 1993 in Wiesbaden, FRG, is an indication, that it has taken place by now:

> " ... the aim (of the new biennial reports) will be to provide elements of prospective analysis which shall be as factual and comprehensive as possible for the European decision-makers" (opus cit. p.22).

> " ... the process of globalization of technology and economy marks the end of national industrial capitalism and generates new forms of ´gouvernance´ of the world economy built upon networks (with variable territorial configurations) of multiterritorial firms, multinationals with the support of local public authorities (the nation-states). This process has long-term structural consequences for the European economies and societies. It will influence the structures of public and private R&D; it will modify the ´rules´ of the game of technological developments and its industrial strategies; it will affect the forms of organisation of the research and the technological markets; in short, the politics of science and of the national and European technology will be highly influenced

by this process" (opus cit., p.16).

The following two FAST research programmes and their outcome seem to me to be most important for the question of Europe´s identity till today:

a ST&CC - Science, Technology and Community Cohesion, and therein especially the strategic role of diversity.

b AT&S - Anthropocentric - or Human Centred - Technology and Systems, especially APS - Anthropocentric Production Systems.

The problem area, where Europe is not only lagging far behind, but where the problems are increasing steeply, can be circumscribed by terms like:

a(i) Regional disparities, unbalanced regional growth, lack of competitiveness of many regions or disregard of the advantages of decentralisation;

a(ii) Lack of cohesion, consensus and convergence between the geographical areas of the Community. Too much willingness to accept uniformity, homogeneity and resemblance, instead of understanding the advantages of diversity.

The task as well as the problem is to turn into advantages what looks like disadvantages. There is no question that, whenever technologies are applied today, they have to be most modern. But knowledge is spreading rapidly now that success is more and more determined by strategic choices, and less and less by fixed assets. At this point the second example comes in: Japanese experiences show that technical instruments are not the only leading edge of productivity and competitiveness and that it is the choice of high-tech combined with work organisation and skills, that counts.[7]

> "AT&S and APS can be regarded as a potential for a resurgence of Europe in general and of European industry/manufacturing in particular. They put new emphasis on human skills and creativity, on which every successful modernisation of European industry and infrastructure depend. Reliance upon technology alone is an inadequate response to the challenges of world markets in the future".[8]
>
> "In times, when competitiveness is driven by the quality customisation of products and delivery times, European diversity suggests a number of possible trajectories of change in manu-facturing, rather than a unilinear path of development or an assumption that there is "one best way" of managing technological change".
>
> "AT&S and APS offer the most important model for European industrial culture. They should also be implemented in the less

industrialised countries in Europe. ... and should give Europe a new industrial attraction in the world".

Conclusion

It is already some years ago that "crisis" was the most talked about scientific paradigm. It was followed by paradigms like perestroika, pollution, chaos or conversion - the fashion changes every year, the problems stay. Or it may be that systems, which are dynamic by definition, are either in constant crisis, or vanish. STD could serve as an example here. Consequently, we have to treat the question in permanence, what size and structure do STD systems need today and what goals ought they to have, in order to:

1 come up to the expectations of those who provide the finances, i.e. to nations and supranational institutions, and
2 to contribute to the advancement of mankind.

It was our aim to look into the question if there is already something like a ESTD, what size and structure it has, or should have, and what questions it treats or should treat in order to merit to be called European.

What was found was that there is no real ESTD-system yet, but that what could develop into such a system is just about right in size, is having structural problems and is quarrelling with the problem how to find the right questions to ask. This now has to be shown in detail.

Part one shows that within the EEC a Europe of big industry is in the making and that political forces with different ideas do have a hard stand. But since it is not yet decided if there is going to be a final winner, industry has to develop ways of its own (ESPRIT versus EUREKA). In view of Europe´s fiercest competitors, the US and Japan, both sides might come to the conclusion that Europe is only going to stay competitive if they work together more harmoniously. But the question remains, what is specifically European in all these endeavours?

In contrast, the European Science Foundation is not only open to the Europe of 12 but to all European states. ESF is closer to science, i.e. less connected to technological developments (for direct industrial or economic use) and to national politics, or, and opposite, more inclined to try true interdisciplinarity, i.e. to assemble natural, technical and social sciences around one table and to tackle ethical questions earnestly and seriously.

Reconsidering the EEC by looking into further programmes with different forces engaged (though there are a whole lot of overlapping), it becomes clear rapidly that there is a wide gap between vision and reality. A programme like DRIVE is inherently fixed to geography and/or national preconditions, i.e. is bound in the typical European tension between national and supranational solutions. And the same holds true for the two outstanding FAST sub-programmes mentioned: Community Cohesion and

Anthropocentric Production Systems.

If we were now to stipulate that Europe would only have a future if "Europe equalled *Campanilismo+Brussels*"[9], what consequences would that have?

One of them could be that Europe would need strong identities. Then the question would be in as far the factual and the future ESTD could develop into such a mean. Or in other words: Are there any signs visible on the horizon for " 'European' social forces with intentions of shaping technical inventions and innovations or even scientific breakthroughs" (paraphrasing "New Technology at the Outset", Meinolf Dierkes and Ute Hoffmann´s new reader with the subtitle: "Social Forces in the Shaping of Technological Innovations", Campus Publ., Frankfurt/M, 1992)?

And with the same intention it could be asked: Could a European Identity develop around STD on Ecological Economy (not determined exclusively by business and industry)?

It would have been surprising, if we had found:

1 that everything Brussels did was fine and a success. Nobody is perfect. But the few examples given show that there are inventions emanating from Brussels, which merit praise and strong further support;

2 that there is no other ESTD than the one originated in Brussels.

The final conclusion is that European diversity is at the same time tremendously frightening and a source of discoveries, ideas, wealth and happiness. This diversity, creativity and innovative potential (scientifically, technologically, ecologically, economically, culturally and in helping to secure world-wide peace), mutually rooted in each other, should also be one of Europe´s Identities.

Notes

1. COSINE - Cooperation for open Systems Interconnection Networking in Europe
2. In order to mention just a few of the EEC-programmes: AIM, BRIDE, COMETT, CORINE, COST, DELTA, DRIVE, ECHO, EPOCH, ERASMUS, ESPRIT, EUREKA, FLAIR, IRTE, MONITOR, PROMISE, PROMETEUS, PROTEAS, REWARD, RISE, SMART, SPRINT, STEP, VALUE
3. FAST = FORECASTING and ASSESSMENT in SCIENCE and TECHNOLOGY

   Random listing of titles listed in the publication-catalogues of the EEC as well as of FAST
   - Legal questions of the future in the EEC
   - Subsidies from Brussels and the future

- European data banks. Eg: CELEX, INFO 92, SCAD, CRONOS, COMEXT,REGIO, EUROCRON, CORDIS
- Worker, employee and professional activities
- Future European Living and Working Conditions (see also the work of the European Foundation for the Improvement of Living and Working Conditions in Dublin)
- Future of young and disadvantaged people, apprenticeship, education, lifelong-learning, etc.. See also the work of CEDEFOP, Berlin
- New work ethics
- Women and computers / Information technology
- Society, communication and information
- Trends in communication and information technology
- Trends in European car industry
- Biological society
- Regional potential of biomass
- Rentability of biomass
- Biotechnology and food
- Biotechnology and the environment
- Biotechnology and the Third World
- Technology, work and professional activities
- The future of food and nutrition, the European food production system and agriculture
- Renewable raw materials
- European priorities in the field of "S&T and Society"
- Specializations of the national industries of the EEC members
- European perspectives for chemistry (scientific, technical as well as for employment)
- European transport problems
- Trends in employment in the service sector
- Innovations in the construction sector and their consequences for employment
- Human language versus machine language
- Impact of computer developments on everyday life
- Creativity, software and computer sciences/technology
- Innovations and new employment
- The future of SMEs (small and medium-sized enterprizes) and employment
- European industry for environment-protection devices
- Local development and new industries
- Electronic housework
- Energy alternatives
- Technical developments and the future of administration
- European life-long learning
- The future of service and repair, from an environmental point of view
- New materials

- Organizations, people and telematics
- New technologies and ecological agriculture
- New crops for semi-arid mediterranean landscapes

4. Institute for Prospectice Studies at Ispra, Italy, Central Bureau for Nuclear Measurements in Geel, Belgium, Institute for Transurani-um Elements, Karlsruhe, Germany, Institute for Advanced Mate-rials, Petten, Netherlands, and Ispra, Institute for Systems Engi-neering and Informatics, Ispra, Environment Institute, Ispra, Institute for Remote Sensing Applications, Ispra and Institute for Safety Technology, Ispra.
5 ESF communications, no.26, april 1992, p.3.
6 ASCEND-Agenda of Science for Environment and Development into the 21st Century. "The objectives of ASCEND 21 were twofold:
   (i) to define the future directions for international science (en-compassing the natural, social, engineering and health sciences) in the field of environment and development for the next 10 to 20 years, and
   (ii) to provide a contribution by the scientific community to the United Nations Conference on Environment and Development (UNCED) in Rio de Janeiro, June 1992".
7. Source: Preparation papers for the forthcoming FAST conference"Science, Technology and Community Cohesion" (Brussels, December 2-4, 1992).
8. The source of this as well as of the two following quotes is: Werner Wobbe, Anthropocentric Production Systems, A Strategic Issue for Europe, EEC/FAST, Brussels, July 1991.
9. Campanilismo means local chauvinism.

13 Technical change in the Czech republic

Jirí Loudin

1

A real content of transformational processes nowadays under way in previous socialist countries is commonly understood as a modernisation. D. Lerner describes modernisation as "the process of social change whereby less developed countries acquire characteristics common to more developed societies". In various specific conditions this process is realised as "Europeanization, Americanisation, or Westernization". Anyway, something like that is in a progress in the post-communist societies of Central and Eastern Europe in those days. The far -reaching societal reconstruction the task of which is to bring a higher quality into all societal activities and sectors and make them more efficient has an obvious model: the Western democracies. Societies in transition are seeking to acquire characteristics common to more advanced part of the world, they are now passing through the processes of modernisation that the latter underwent in the past.

Modernisation basically means dissolving traditional forms of life and refers - according to J. Habermas - "to a bundle of processes that are cumulative and mutually reinforcing: to the formation of capital and the mobilisation of resources; to the development of the forces of production and the increase in the productivity of labour; to the establishment of centralised political power and the formation of national identities; to the proliferation of rights of political participation, of urban forms of life, and of formal schooling; to the secularisation of values and norms, and so on." For the society of the second half of 20th century, to be modern means to be based on "Grundinstitutionen der Konkurrenzdemokratie, der Marktwirtschaft und der Wohlstandsgesellschaft mit Massenkonsum und Wohlfahrtsstaat"(3, p. 34).

The post-communist societies of Central and Eastern Europe can be hardly identified as fully traditional ones. In their functioning in last decades, they were traditional in specific form, perhaps even in their substance, due to an

absence of two basic corner-stones of modern societies: market economy and democracy. In other respect although, they were rather modern and besides that the conditions were quite different regarding time and space. In some countries, traditional societal structures were established or substantially strengthened by force after Second World War.
The post-communist societies have after decades of social (enforced) experiments decided definitely to abandon a search for "the third way" and accept a standard social order according to Western model. This process is widely accepted by the public and there is no menace of such troubles in accepting modernisation as was the case in some developing societies, where was done an attempt to introduce Western-like modernisation not considering a substantial cultural difference. Such problems are here quite improbable at least for the first period of modernisation, when the dominant idea is to liberate himself from the past, modernisation equals basically liberation and the new identity of transforming societies with new setting of real interests is not ripe enough as to give rise to possible conflicts of interests with the modernises at all.

2

The undoubtful success of the West in material, cultural as well as spiritual respects is closely connected with a high level of advancement of science and technology, it is by its nature a technology society (understanding technology as an applied science is symptomatic for modern society). Modern science and technology represent together with a market economy and political democracy three mutually intertwined corner-stones of this civilisation. It is not by a chance that the concept of modernisation is linked with the concept of "Occidental rationalism" as described by M.Weber. According to him the rationality is based on science which is able to provide us with knowledge how to master conditions of life as well as human behaviour by calculation.

It is science and its application in technology what strengthens autonomy and freedom of man. In last two centuries since the middle of 18th century the cultural development of the West is characterised by the ideas of Enlightenment and Liberalism with their hopes in a progress consisting of combination of science and democracy. According to some radical liberal theories, as a normative model of the whole society should serve science community with its guarantees of free, unhindered discussion, public evaluation of the various issues, promotion of criticism, argumentative, justifiable reasoning and acting.

The post-war development of interrelations among science,technology,and society is instructive. The era of the fruitful, harmonic co-existence of science,technology and society and optimistic faith in unlimited potential of science and technology culminated evidently in the period of post-war reconstruction and growth. This period is characterised by extraordinarily high degree of involvement of science and technology into social end economic development and its entirely prevailing positive acceptance by the public. An impression was arising that science and technology is able to solve every social and economic problem included those which were caused by the social application of some specific technology. Then the solution is to find and apply even more perfect technology. The only problem to be solved is to have at disposal enough capital and qualified work force.

This situation has entirely changed during the sixties owing to appearances of the serious negative effects of science and technology: pollution, deterioration of social and professional structure, a menace of misusing the new information technologies for manipulation with people. A failure appeared to be the concept of science and technology development as a self-sustaining process, without setting social priorities.The criticism toward science and technology which was in the next period very strong reflects all the troubles, traps and obstacles caused by science and technology. However, at the same time, it is a part of wide societal changes. Society faces the value crisis and behaves to science and technology more sensibly. Science and technology functions here as a representative of old societal and value structure the substance of which is consumerism.

The changes in value structures which were arising and spreading in the sixties and seventies are mostly described as the transition from Materialism to Post-Materialism and their nature may be characterised as "shifting from an overwhelming emphasis on material well-being and physical security toward greater emphasis on the quality of life"(4, p.3). While "material values reflect a relatively strong attachment to maintaining order and preserving economic gains. Post-Materialist value emphasise individual self-expression and achieving a more participant, less hierarchy society"(4, p.179). The efforts to get the social effects of science and technology under control led to the establishing of such institutions as technology assessment. In previous socialist countries, the system of science and technology developed in quite different conditions. We wouldn't find here the absence of science and technology as it should be the case in a traditional type of society. On the contrary, support of technical change based on science advancement was the important part of the official ideology and policy. Nevertheless, at an international level, a gap was appearing and even increasing in those areas, too. Science and technology acted in the same economic and political conditions as other institutions and they may have missed competitive economy and democratic order more than the latter. Science and technology did not function as an independent agent which is connected by many equal, horizontal links with development of people and institutions. They represented a functional, instrumental part of whole-society plan which was hierarchically structured. Regarding the knowledge potential of science, it was utilised - namely in the form of social sciences - as a theoretical basis of so called "scientific administration of society" and hence it was used for a manipulation with people.

It was not an open society. It was society with supremacy of politics over the other spheres of society and with monopoly of one party in politics. In this respect, it was traditional society. System theory conclude that the modern society, unlike traditional one, don't contain any authoritarian centre which would play a dominant role in whole society. Modern society is functionally differentiated and its subsystems such as economy, state, education, science etc. enjoy large autonomy. They are linked together with functional relationships and create the equilibrium by balancing one another. None of the subsystems can seize control over the others. On the other side, the traditional societies are structured hierarchically and are steered from the top of a hierarchy. In the totalitarian regimes it is a political system and his representatives - state, political parties and leaders - that execute supremacy in all the fields of society.

What was missed most was the autonomy, intellectual freedom and stimulating economic incentives. The results of science and technology were urgently needed and those spheres got real economic and political support, indeed. However, at the same time, those were kept a tight hand on full of the ideological prejudices and economic contra-incentives. Each techno-economic solution became a political issue, too. The research system could not develop the functional relationships with other subsystems of society. The central departmental command system caused that even the mutual relationships among individual segments of research system were hampered. Thus the barriers among basic and industrial research and development arose, poor horizontal ties among specific research branches as well individual research institutes existed.

The official ideology represented some sort of technological determinism with the concept of science as "direct productive force", subordination of science and industrial research to the needs of production. At the same time, the central command system with prevailing vertical links factually isolated research and production each other. Besides that, the production was backward, oriented on not demanding markets. In that way, the backwardness of production threw to the same position the research system , too.

Pre-revolutionary regimes appeared in a difficult position. Their ambition was to raise the performance substantially,compete with the most advanced societies and close the gap. They accepted not only their technologies and know-how but to some extent necessarily their interest and value structures, too. Simultaneously, they intended to keep their own identity, which was, however, in contradiction with the former. This cleavage was widening as the development of modern technology based increasingly on certain values which could not fully develop in social and political conditions of that time. Those values were especially self-realisation, responsibility, autonomy, participation, communication. It was no more enough to possess specific professional skills and knowledge, the integration of knowledge and values (orientation knowledge) gained on importance, and the very professional ability was becoming more and more tied with the specific value orientation. In spite of the unfavourable conditions in society, those new trends in value orientation were expanding in society, especially in some strata, but there were too many obstacles here and decisive power structures had not a positive attitude to it.

A thesis of the ambivalence of value structures in then so called socialist countries in the eighties can be well documented on the results of several surveys the task of which was to investigate various social aspects of technical change and which were conducted in the middle and second half of the eighties in Czechoslovakia by the research group of the Institute for Philosophy and Sociology of Czechoslovak Academy of Sciences.

In a key area of job values, a high significance was laid on self-realisation, autonomy in work; work satisfaction and enjoyment got a very high score. However , all that occurs often linked together with a low willingness to take risks and responsibility, do and promote changes. These attitudes express rather the inclination to being passive and static.

The survey found out a pretty firmness in standpoints relating to the environmental aspects of economic and technical development. The vast majority of population being interviewed (93 per cent) would agree with the reduction of production dynamics if continuing increase of production and consumption would cause further deterioration of the environment. Considering disastrous environmental situation in Czechoslovakia it might not be surprising, however, it is remarkable that people are aware of such a high price for the desired solution and would probably accept it even in conditions of greatly unsatisfied needs.

We have already mentioned the clear positive acceptance of technology by the public of Western democracies in first decades of post-war reconstruction and the following decline of such agreeing attitudes. In the former Czechoslovakia, those positive attitudes had been preserving until recently and the situation is not quite different now. In Germany in a series of surveys were people repeatedly asked whether, in their opinion, technology represents for mankind rather blessing than damnation. In 1966, 72 per cent voted for blessing; 20 years later, in 1986, only 41 per cent of people were of the same opinion (5, p.288).

In 1984 in Czechoslovakia, the respondents were given a question, whether they consider the effects of technology for people rather positive or negative (what is, in my opinion, a comparable question with the former). Sixty seven per cent declared the effects for positive. Hence the result is close to attitudes which were found out in Germany in 1966.

It seems that the post-materialistic upheaval [a growth of criticism toward technology is commonly considered as a part of it] has not yet been accomplished, at least in our country. The results of more recent surveys speak for such contention. According to some surveys, what are people most expecting from new political conditions is a rise of material standard of life, a higher level of satisfaction of material needs. No wonder considering the existing gap and new emerging possibilities in that area. Will this fact mean the positive impulse for the development of technology or will dominate destructive impact of such interest and value structure on technology in terms of underestimating risks and dangers?

3

Several years of effort at modernisation in post-communist countries justify the question to be asked how the technical change participates in societal transformation, and what changes appeared in development of science and technology in new social conditions. Is the transformational task in that sphere, that means to raise the innovation capacity of society nearer than earlier? It is possible to sketch some model situation which reflect problems the societies in transition are coping with. Basically, there are two different ways of coping with technology in such circumstances. The neo-liberal concept in its extreme form doubt a need of the articulate technology policy. It considers technology as one important production factor among others that is not so extraordinary as to require specific respect, eventually specific policy. According to this opinion, the best guarantee for a sound technology development is free, unhindered functioning of market forces. This conceptual and political stream devaluates whatever attempts at formulating a long-term strategy as an ungrounded pride of reason that has always failed. The breakdown of socialist system is taken as a proof for such statement.

The second conception claims also a major role of a market however it lays great importance on a long-term dimension of technology development where should be engaged to great extent the state and the public . It is seeking to elaborate goal-oriented, selective strategies that would create the positive prerequisites and incentives for technology development and thus would help prepare high-quality inputs into the markets of future. Supporting promising trends in technology, setting priorities, elaboration of educational programmes, strategy of involvement into the international technology development are all parts of this endeavour.

However, the real situation is never solved in "clean", model way, it is composed of the elements of both methods. For many reasons, in many post-communist countries was chosen neo-liberal variant of economic transformation and therefore those neo-liberal elements dominate - at least in this first period of transformation - also in an area of science and technology development. The reasons for rejection of whatever form of state interventionism are various: it works here the historical experience of last decades` hypertrophy of state interventionism, against it are the doctrinal (nothing can be wiser than the market) as well the financial arguments (shortage of money) and the role play the political circumstances (state is represented by not yet transformed bureaucracy and each state activity means the strengthening of it).

Although the strategies of both economic transformation and technical change may differ in emphases (which are changing in time - if we can refer to Schumpeters concept of "creative destruction", so with some simplification, in first phase the destruction is dominating and then is coming time for creativity) and in timing (fast and gradual transition) the crucial real problems are basically the same and are related to transition to market economy.

The opening of society as well economy does mean the end of political restrictions and interest separation of science and technology from market economy and production. Science and technology is exposed to new pressures and demands, they are stepping in new, more articulated links, which have not the command nature but have very high relevance.

The structural interdependence of both research and production system which were pushed back is being restored now: the economy was deprived of natural ties with world economy and knowledge potential of world research and owing to that isolation the society built up his own , over-dimensioned, comprehensive system of fundamental and research system, which covered all the spheres of reality and the task of which was to substitute insufficient information flows and exchanges with the world.

The restructuralization of science and technology is realised under two kinds of pressure: production can in far higher extent receive technology and information from abroad, and, simultaneously, the very structure of production is changing as a consequence of the opening of the economics outwardly and hence the demands toward science and technology are changing, too. The urgent problem is how to put together science and technology development with the entrepreneurship. The decisive factor is to what extent is the transforming economy able to accept and promote the technology innovation as a core of entrepreneurship. On the other side, it is in interest of research system to reflect the fact that an economic growth is rather the result of the diffusion of knowledge and its entrepreneurial utilisation than the knowledge production itself. Anyhow, the implementation process by its nature cross the technical and economical dimensions toward social, cultural, and educational aspects.

The abolition of political limits raises the question of democratic control of science and technology by the public and the nature of participation of state in science and technology policy. In this area, we can observe the trend to distribute responsibilities among a higher number of subjects of civic society. In the turbulent period of societal transformation it is very sensitive problem: there are some groups and civic initiatives which are very active, on the other side it must be taken into account an apathy of some part of the public which follows from the unexpected difficulties of the transformation and some role plays also the passive consumerism which was described above.

With the opening of economy the possibility is enhancing to acquire new technologies with the help of technology transfer. If post-communist countries want to close the technological backwardness-gap, they need necessarily technology assistance of world technology leaders. It is again the question arising whether that task is fully resolvable by means of a spontaneous activity of market forces or it should be in some way oriented and structured. Some kinds of technology transfer may make from the recipient country a huge assembly hall where an advantage of cheaper workforce is valorised and the learning process is minimal.

If the recipient country wants to play more active role and to gain an access to new skills and knowledge, the favourable conditions for that must be created on the side of importing country. That means to maintain a high quality and appropriate structure and orientation of education and to support inland science and technology research - structured in priorities but still wide enough to be able to catch up the new, sometimes unexpected innovations from outside. Technology transfer is at the same time transfer of culture and social structure. Now, to what extent must technology to be transferred to Central and Eastern Europe be adapted in terms of cultural and social compatibility? What is the real difference in culture, mentality, social structure?

Although the post-communist societies are manifesting their affinity to the Western cultural and political tradition, those cultural norms and values remain largely at the level of projects, programs, ideas, and are not sufficiently interiorated and manifested in social activities, work ethics, habits and norms of personal contacts and communication. Central Europe is a region of mixed culture and hence there is a high-degree adaptation capacity available here. It is obvious that it would be very short-sighted to assume blindly the foreign models. In reality, at least so far, technology meet with great obstacles. State has not enough money, domestic capital is still too weak and foreign one too cautious.

The attention is focused on the research system and policy. The research system was, namely for reasons of international isolation strongly over-dimensioned and is undergoing the radical changes. The research system, having liberated itself from the bonds of command system is looking for a new, more appropriate social basis with more structured and developed ties with economics, culture, education, and values. The trouble is that neither R&D nor the other societal institutions have so far gained desired autonomy and independence. In the transition period full of uncertainties and shatterings, they are still weak and concentrate on their own survival and strengthening not having enough energy to cooperate mutually and promote each other.

In dramatically changing economic and social conditions is urgently felt the need to re-conceptualise a role of R&D in society and economy. The problems which arise in this process can be described on the example of transformation process of the research system of the Czech republic.

In first post-revolutionary period, the main task was to get rid of all political constraints and enable the people to conduct free research. The program was based on the concepts of democracy, plurality, and competitiveness. It can be said that on the basis of applied changes of legislative, organisational, and personal character, those principles are applied in both resolving the research tasks and acquiring and distributing the financial resources. Immediately after a change in political power decentralisation and democratisation in science and research was realised. Those were followed by the changes in a legislation; the new norms regulating activities of higher education institutions, Academy of sciences, and the forms of state support to science and technology were introduced. There were established the new governmental bodies, as for example, Council for science activity and technology development.

The efforts to develop and promote innovation policy in our country meet great obstacles which correspond with fundamental nature of society transformation and its science and technology potential. The initial stage of the development of market forces and mechanisms, the economic stagnation with its lack of financial resources have not so far provided the research system with needed dynamic basis. There is a low demand for new research and development solutions. The companies that are still non-privatised have no clear idea of the long-term entrepreneurial intentions as well prospective production structure and his main idea is to survive for a short next period. The technological innovation has not been perceived as a core of entrepreneurship. In such conditions is it very difficult to set national priorities in research.

That applies especially for the sphere of applied and industrial research which is tightly connected with the production. Industrial research is in a real crisis; a number of researchers is constantly diminishing, some institutions are being dissolved, the other are being privatised. A reply on the crucial question what kind of technology policy to be pursued in this transitory period when neither former command system nor future market one really exists sound as follows: to accelerate the transition, go through the dangerous time as fast as possible. The best technology policy should be to carry out deep macro-economic changes, not to construct priorities, means, and stimulating mechanisms specific for this sphere and period. At the same time, spontaneous rise and development of supportive networks to which belong science parks, incubators etc. go on slowly and with many obstacles.

Considering a poor condition of our industry and industrial research, there is no wonder that the attention is focused on basic research. Science is appreciated as a common, general value which is to be preserved and cultivated as a thesaurus of vast cultural and intellectual potential which is unavoidable for a long-term technology development. This fact is (mis)used by some scientists for underassessment of technological knowledge whose specific nature and identity are denied (though, in fact, the contrary is true; the technological knowledge is unique and is transferable with great difficulties; it is no freely accessible on the market).

The project of transformation of science includes three key dimensions, i.e. institutional, personal, and cognitive changes. While changes of institutional as well as personal kind were carried out relatively fast, to increase a quality of cognitive activities is more comprehensive task which can't be realised at once. So far, the transformation of science is more understood and realised as a process of institutional transition in new political and economic conditions than the real increase of investigating on the level of modern research systems integrated into international intellectual networks. The trends of new programs and projects typical for international research in recent decades are identified and absorbed only to a limited extent.

The experiences from the transformation of the research system in the Czech republic show that perhaps the first, initial phase of transition is closing now, in which the main task was to overcome the inertia of old mechanisms, identify nature and structure of new arising problems. Now is emerging the need to constitute a new concept of research and innovation policy and harmonise it with a course of societal and economic transformation. It is a way of conduct which is appropriate for the modern society.

References

Habermas, J (1987), *The Philosophical Discourse of Modernity*, Polity Press. Close to it is Lerner's description of modernity: "a degree of self-sustaining growth in the economy, or at least growth sufficient to increase both production and consumption regularly; a measure of public participation in the polity - or at least democratic representation in defining and choosing policy alternatives; a diffusion of secular-rational norms in the culture - understood approximately in Weberian-Parsonian terms; an increment of mobility in the society - understood as personal freedom of physical, social, and psychic movement; corresponding transformation in the model personality that equips individuals to function effectively in a social order."

Inglehart, R (1977), *The Silent Revolution, Changing Values and Political Styles Among Western Publics*, Princeton-New Jersey.

Lerner, D (1969), *Modernisation; International Encyclopaedia of Social Sciences.*

Noelle-Neumann, E (1988), 'Medienwirkung und Technikakzeptanz', in W. Suess, K. Schroeder (hrsg.), *Technik und Zukunft. Neue Technologien und ihre Bedeutung fuer die Gesellschaft*, Westdeutscher Verlag, Opladen.

Zapf, W (1990), 'Modernisierung und Modernisierungstheorien', in *Die Modernisierung moderner Gesellschaften, Verhandlungen des 25. Deutschen Soziologentages in Frankfurt am Main 1990*, Campus Verlag, Frankfurt/New York.

14 Changes versus legacy. Science and technology policy in Bulgaria

A. Yossifow

Science and technology policy research most often deals with one and the same type of elements - financing, potential, subject orientation, influence groups - their dynamics and the social consequences of their activity. It is more difficult (and has been done less frequently) to analyse science policy within the context of a particular society's political norms (which have been established outside of science), within the predominating ideology, cultural traditions and, what is more important, according to the concept and image of science current in the social consciousness. However, this is the only means of explaining certain phenomena in Bulgarian science and technology which are part of the past, and hopefully, on the way to being eliminated.

The change in Bulgaria was faced with a hard legacy in the area of policy relating to science and technology (just like in any other area). The following factors determine the state of affairs in this area:

Heavy bureaucratic structures, typical of totalitarianism, unified in all spheres of human activity, and facilitating total control. Plato in his day, followed by Aristotle, defined the three types of social order: monocracy - rule by a single individual, a Monarch; aristocracy - government by the best people or in the name of the best goals; and ....(known as democracy today) - government by majority. Again it was Plato and Aristotle who warned of their possible degradation into tyranny, oligarchy, rule by the mob. Bulgarian totalitarianism managed to combine all three types of degradation.

Forms of organisation of science and technology management which are the exact image of the forms created in another powerful ex-state, and applied, of course, according to our country's means, scale, and our peripheral conditions. In contrast to the Citadel, this process was not accompanied by the masses' spontaneous and honest faith in the ideological foundation in Bulgaria. This made the practical application of these forms more cynical from a moral point of view, and less efficient from the point of view of practice. Even the meaning of social organisation differs in the different cultures, and there are no "exemplary" forms which could be borrowed.

Phenomenal respect for knowledge and its bearers which fits into the patterns of our cultural traditions. I shall mention several sociological indicators: eighty four percent of high school leavers wish to become university students. The percentage is almost the same for university students intending to devote themselves to the carrying out of research once they have graduated (data gathered from a 1989 survey). The total amount of volumes of scientific literature and fiction in print is comparable to the amount in England, France, Italy, and Spain, and book shops quickly sell out of stock. The average book collection for a family residing in an urban area totals around 200 volumes. Thirty five percent of the members of parliament in the National Assembly are currently doing research work, or once had been engaged in scientific work. A survey on nuclear power indicates that the credit given to scholars from the Bulgarian Academy of Science is greater by between eighteen and forty points than that which is given to the Government, the Parliament, or foreign experts. All of which creates a favourable psychological atmosphere for innovation.

A predominant conception of science and technology different from the one shared by the world scientific community. This conception involves priority for neither independent knowledge, nor expertise, neither cultural monitoring, nor training of specialists. In the governing of people's esteem science has, first of all, representative functions, both on a national and on an international level, and its aim is to attach legitimacy to other decisions and activities. Science has deliberately been turned into a professional niche for both people who follow their vocation (but have limited ambitions) and people who have already proven that they lack competence in other areas, but however, they enjoy some privileges and deserve a relative social stability. The replacement of basic by secondary functions and the concept of science as a nomenclature reserve are not only in someone's mind. Twice in our recent history - in the late 1950s and in the middle 1970s - some not quite responsible and unselfish scientific leaders promised fast and imposing technological results and ruined several national programmes. As a consequence, the politicians regularly ignore science (although this is not plainly expressed) when important decision making is concerned. In areas where science is irreplaceable (national security, power engineering, social conceptions, etc.) the principle followed is to rely upon the "brotherly" help from abroad.

These factors have a determining effect on our results in science and technology policy, and are worth further consideration in this paper.

The scientific institutions

Totalitarian principles have imposed a strict hierarchical order which enables universal control. This implies centralisation and a relevant development of the bureaucratic apparatus. Each research unit has its "superior level" which usurps some of the responsibilities of the lower levels without having complete authority for final decision making. The Bulgarian Academy of Sciences, for example, is a research institution that has no analogue outside of Eastern Europe, but is suitable for realising the centralization in question, and is constructed in such a way that it could serve these purposes. Research plans, international relations, fund allocation, professional career paths, as well as all scientific activities are subject to control from "higher quarters".

It is true that science and technology policy intervenes at a certain stage by stimulating and canalising ideas and professional practice, having emerged and been tested in parallel structures. However, structures parallel to the official ones are rarely created in centralized science, and, if at all, they are usually short lived. Bureaucracy does not tolerate variety. An indicative fact for example is that eighty percent of the research programmes are planned for a five year period, irrespective of the complexity of the task. At the same time, post graduate studies have a fixed period of three years for all disciplines, a time period which does not take different conditions, themes, and the creative abilities of the post graduate students.

Characterized by these features, the institution such as it currently stands is not a means of raising the work efficiency of its members, but rather an instrument for oligarchic control.

Funds and fund allocation

The state budget is currently the only source of financing techno-scientific progress. It covers around ninety-four percent of the total expenditure either directly or through other, also state, institutions. Until recently there were no research foundations in Bulgaria, nor were there any autonomous and sufficiently powerful economic forces which could become sponsors of science. In the last few years, such foundations were created, but only operate within a limited thematic and staff perimeter. Grants and financial aid coming from abroad are also centrally allocated. All of which accounts for the priority financing of those research groups which are in direct contact with the central administration. Sometimes these projects used to be extremely expensive, over ambitious, incredible, and proved their own absurdity within several years.

In order to illustrate this, several examples are particularly noteworthy:

(i) A Bulgarian overpowerful computer, designed and produced in two models, using imported constituent components, by a team of five hundred people and directed by an ex-minister. As a result, the computer was exhibited at two fairs.

(ii) A programme in suggestopaedia which experimented in learning a foreign language in a few days (this experimental institute survived for eight years).

(iii) A new method of oil prospecting in the Libyan desert (costing the Bulgarian state two hundred million dollars).

(iv) A programme for the development of semi-conductor electronics whose centre was located in the former head of state's village.

(v) A project which looked into the possibility of using water as a fuel for internal combustion engines!

Such projects are preconditions for idleness creating a feeling of permissiveness in those working on them.

Setting aside such paradoxical exceptions, the principles of financing remain subject to the above factors. Two point five percent of the national budget allocated for science is unevenly distributed between the institutes and the universities (a seven to three ratio respectively). At the same time, the proportion of the wage funds increased at the expense of the funds allocated for equipment, information, and immediate research work. Thus, the professional niche is provided for while the social function of science is desperately underestimated.

Staff recruitment

The system of scientific and technical staff recruitment is crucial for the normal reproduction of the professional community. There is no system of specialist selection currently in Bulgaria. The academic "pipe" blows up everyone who has the desire and tenacity of staying in the profession to the rank of professor. Dismissal is incredible and only possible at the last resort (if, for example, a coup d'état is being attempted).

Selection procedures upon entrance to the profession are carried out through tests. These being something like an armed confidence in university education. However, since competitions do not allow for creative ability and ability for risk to be taken into account, erudition and memory come to the forefront. It is as if when selecting an airplane pilot, the contender who best knows the biography of the Wright brothers, or the timetable of the local airline is chosen.

Bureaucracy has introduced nine hierarchical levels into the scientists' profession. Advancing to the higher levels involves an element of secondary selection. Here again, external factors and the relations with the centre have an impact. The attainment of an academic rank, for example, requires in addition to scientific publications, the issuing of nine administrative documents.

At a certain stage, discussions are held and the candidates are even put to the vote of the scientific council. Their voting, however, is endorsed by a supreme committee composed of specialists in different fields of science. These specialists neither know the candidate, nor are aquatinted with his/her scientific production. They are only familiar with the preliminary intentions and make their final decisions in accordance with these intentions. Data from our survey shows that corrections have been made by the committee in eleven percent of the cases, and these corrections have always followed the previously indicated direction.

There is another mechanism known in Bulgaria by the slang word "parachuting". Some of the scholars have been appointed, rather than elected, by the men of the day. They are usually people who do not possess any particular scientific qualities. They represent fifty percent of the current research staff according to some evaluations.

With this in mind, science is a prestigious and attractive occupation for young people. In Bulgaria, no business has existed neither theoretically nor in practice for many years, art only existed theoretically and politicians were selected according to other criteria. As a result, almost the entire contingent of young people whose abilities are higher than the average turned to science. Thus, the high creative potential of the other fifty percent is objectively predetermined. A fact which may well be surprising to the foreigner who comes into contact with Bulgarian scholars, for he/she finds a great difference between them as far as their abilities and professional training are concerned.

The role of the university

The university as a cultural institution bears much significance for society. Not only does it organise knowledge in order to render it in a synthesized form, but also in one sense it disciplines thinking, creates an orientation to a learned and prejudice-free commitment to problems. The university also has the advantage of being in constant contact with the young generation. In advanced science, the functions of research and education are inseparable, as well as being mutually determined.

Universities have greater resistance to total control. Their struggle for autonomy gives rise to permanent problems. The government is fully aware that it is more dangerous to underestimate education than research. The former serves the dissemination of ideologies. The risk accompanying the intellectual adventure inherent in genuine research is assiduously avoided with students. The outside world (policy) has to be cut off. In the light of student strikes, we can judge that the strategy has not been quite effective, but the tendency of separating science from education is preserved. The two functions are being performed by different people in different institutions, and artificial barriers are being set in front of the national transition of specialists from one type of activity to another.

The problems in each of these fields are numerous, and it seems that our science policy is characterized by a greater number of vices than virtues. The rational organisation of scientific life is extremely difficult, especially under the conditions of the present crisis and the existing inertia. Additional efforts are necessary in order to follow the present course of the scientific progress - cultural liberalization, informatization, and internationalization. The challenge is awaited with baited breath.

15 Adaptation to European union: The adjustment process in Greece

Zissis Papadimitriou

Resolutions made by the political leaders of the European Community member states at the conference in Maastricht brought the vision of united Europe closer to the minds of European people. However, the realisation of this vision will depend, in the last instance, on the ability of the European Community to face successfully the developmental problems of the small, less industrialised member states of the Community, such as Greece, Ireland and Portugal in the long term.

The possible scenario of a "two speed Europe" has already been outlined, and points out that the southern EC countries in particular are not able to follow the economic, technological and social evolution which is underway in most of the more advanced industrialized countries of the EC. Greece is a typical example of a member country which faces the danger of marginalisation while confronting serious problems of "adjustment" to the Community standards because of the structural and organisational weaknesses of the Greek economy as well as its technological backwardness. If the association treaty with the European Community in 1961 set the new course for the integration of Greece into the international division of labour, then the accession to the EC in 1981 marks a new phase in the country's economic, social and political development. Both the realisation of a uniform European domestic market at the beginning of 1993, and European monetary union, which has been fixed (albeit provisionally) for 1999 at the latest, imply a great challenge for Greece. The country will have to adjust itself to EC conditions if it wants to meet the requirements resulting from the common European domestic market and monetary union. The success of this adjustment process will largely determine whether Greece will find its way into the unified Europe. In doing so, the country will have to solve a series of problems which are dealt with in the following publications concerning Greece and the European Community: T. Yannitsis, Accession to the EC and effects on industry and trade, ed. Foundation of Mediterranean Studies (FMS), Athens 1988 (in Greek), N. Maravejas, Accession of Greece to the

EC: Effects on the agricultural sector, 1981-1985, ed. FMS, Athens 1988 (in Greek), L. Tsipouri, Accession to the European Community and Greek Technology Policies: The experience 1981-1986, ed. FMS, Athens 1989 (in Greek), P. Kasakos and K. Stephanou (eds.), Greece in the European Community: The first five years - Trends, problems, perspectives, ed. Sakkoula Publications, Athens 1987 (in Greek), K. D. Christou and A.H. Sarris, The impact on Greek agriculture from membership in the European Economic Community, European Economic Review, vol. 14(1980), pp. 159-188, Commission of the European Communities, Report on the competitiveness of the Greek economy, Brussels, II/ 485/86 - EN, Rev. 6, W. Hummen, Greece and the European Community: Problems and prospects of the Greek small and medium industry - Results of an empirical survey, German development Institute, Berlin 1978.

The ability of the Greek economy to adjust to the European standards is of great importance even for the future of the country. The "new order" which resulted from the collapse of the Soviet Union and its allies, led to the geographical resurgence of Greece in the Balkans and in the Middle East as well. Greece could play an even more important role in the context of Community's external economic relations (trade, capital investment, finance activities etc.), especially concerning the former socialistic countries of the Balkan Peninsula. The "conditio sine qua non" for such a development is the economic, technological and social harmonisation of Greece within the European Community.

1 The development of the Greek economy in the post-war period

An overview

In an attempt to systematise the course of economic development of Greece in the post-war period, roughly three phases can be discerned:

a) from 1953 until 1960,

b) from 1960 until 1974, and

c) from 1974 up to today.

With reference to the definition of periods of the economic development of Greece, see K. Vaitsos/T. Yannitsis, Technological transformation and economic development - The Greek experience and international perspectives, ed. Gutenberg, Athens 1987, p. 15 f(in Greek).

For the other European countries, the second World War had finished by May 1945, but in Greece it continued as a civil war from 1946 until 1949. Following the end of the civil war, the country found itself in a desperate state. Only in 1953 did a turning point appear when the reconstruction process of the Greek economy was launched. In this year, the government took several measures which were to contribute to the stabilisation of the economy. The most important ones were: cutting state expenditure, reducing imports, intensifying control over the awarding of bank credits, freezing wages and salaries below the growth rate of productivity, and last but not least the devaluation of the Drachma by 50%. See A. A. Kintis, Development of Greek Industry- Accomplishments - problems - perspectives, ed. Gutenberg, Athens 1982, p.28 (in Greek).

In the course of the reconstruction process, the war damages could largely be repaired and the extension of an infrastructure necessary for the industrialisation of the country, particularly the transport and communication system, was initiated. At the same time, the production potential was doubled. But inspite of these successes, the Greek economy remained a backward rural economy, with the exception of navigation which occupied an important position on the international level. Thus, neither the agricultural nor the industrial sectors showed any significant changes in the resulting structure of production . As to agricultural production, almost 80% of it consisted of but a few products, namely tobacco, raisins, cotton, oil, olives, and wine. The structure of production in the manufacturing industry, too, showed little differentiation. Two thirds of its products were used to cover domestic demands. See K. Vaitsos and T. Yannitsis, op. cit., p. 16. Most of the production was used for domestic purposes. Only at the beginning of the sixties was there an onset of a certain qualitative differentiation of production when Greek entrepreneurs became interested in the extension of a more export oriented economy. In the fifties, the institutional and infrastructural prerequisites were set up for the actual growth process of the Greek economy which was not to take place before the following period of 1960-1973. During this period, the transformation of the Greek economy from an up to that point rural economy into a semi-industrialised one was accomplished, Ibid., p. 18.

Thus, in the 1961-1971 period, manufacturing industry production increased by 175%. Its contribution to Greek foreign trade went up from 10% in 1960 to 41% in 1973. A large share of the economic growth was similarly due to the tertiary sector; the volume of which increased by 98% between 1961 and 1971, Ibid., p. 24 f. and p. 46.

Agriculture, too, was able to register successes in production. Thus, besides traditional cultivation, new crops (rice, sugar-beet, vegetables, fruit etc.) were introduced which decisively contributed to the qualitative improvement of agricultural production. With the aid of new cultivation methods and the use of fertilisers and plant protection products, agricultural yields improved considerably, Ibid.

The average growth rate of the gross national product (GNP) was about 7.2% (1960-1973) and thus was the highest that Greece achieved in the post-war period. Concerning the increase in labour productivity, Greece ranked first among western European OECD countries with an average growth rate of 8.1% (1965-1973). The per capita income of the Greek population also improved considerably. In fixed prices of 1970, namely from 574 US dollars in 1960 to 1133 dollars in 1970 and 1966 dollars in 1985, Ibid., p. 19.

In 1974, the dynamic growth progress came to a standstill. Labour productivity growth rate declined to -0.8%, levelling out around 1.7% between 1975 and 1981. Since then it has stayed on this level with only minor fluctuations. See OECD, Productivity in Industry. Prospects and Politics, Paris.

From that time on, the Greek economy has remained in a deep structural crisis. At the same time a slow reduction of the industrial system has taken place. The process of deinvestment and de-industrialisation manifests itself above all in a shift of investment activities from the industrial to the tertiary sector, and here again primarily into non-productive activities (money transactions etc.). See K. Vergopoulos, The de-development today: An essay on the stagnation dynamics in Southern Europe, ed. Exandas, Athens 1986, p. 174 (in Greek) as well as K. Vaitsos and T. Yannitsis, op. cit., 22 and T. Yannitsis, Greek Industry - Development and crisis, ed. Gutenberg, Athens 1985, p. 374 (in Greek).

The crisis became visibly aggravated following the accession of Greece to the EC in 1981, especially since the country's economy has been more and more strongly exposed to competitive pressure from the other EC member states as well as from non-EC countries. Instead of adapting to EC-standards, the Greek economy decreased constantly during the 1981-1992 period. See Commission of the European Communities, "European Economy", com. no. 50, December 1991.

If the efficiency of the economy could not be improved by restructuring and technological modernisation of the production, the marketing difficulties of Greek products might well be aggravated after the completion of the EC domestic market. The crisis in which Greece finds itself at present is revealed inter alia by a continuously growing deficit of trade and in an immense national foreign debt. The deficit of trade amounted to 6.950 billion dollars in 1987. Today, Greece is one of the EC countries with the highest foreign debt. See G. Karabelias, State and Society in Greece after the military dictatorship, 1974-1988, Athens 1989, p.288 f.(in Greek).

2 Structural problems of Greek industry

According to statistical surveys, the industrial sector numbered 144,463 industrial and manufacturing enterprises with 684,145 employees altogether in the middle of the eighties. About 99% of these enterprises are small and medium sized (1-49 people) comprising almost 64% of the employees in the manufacturing industry. See National Statistical Service of Greece, Stastical Yearbook of Greece 1987, Athens 1988, p. 226 f.

The highest increase was shown by the number of small enterprises with 1-4 people. This development is mainly due to the fact that, as a consequence of the enduring crisis, many enterprises had to dismiss personnel or even to shut down, whereas a tendency for small units prevailed among newly founded enterprises. See Centre of Planning and Economic Research (KEPE), Small and Medium sized Enterprises in the Manufacturing Industry, Athens 1988, p. 37 (in Greek).

This seems to be explained by the greater flexibility potential shown by small and medium sized enterprises, as compared to large ones, during the crisis, Ibid., and A. Lyberaki, Small firms and flexible specialisation in Greek Industry, PhD. thesis at the University of Sussex (IDS), 1988, p. 168 f. Concerning the efficiency of the various sizes of enterprises in Greek Industry, see also D. Stamatopoulos, Industrialisation and strategy of domination in Greece, ed. Foundation of Mediterranean Studies, Athens 1989, p. 182 f (in Greek).

The employment structure, too, of the Greek economy happens to correspond with the preponderance of small and medium sized enterprises. The Greek employment structure is characterised by comparatively high number of self-employed people and professionals. The share of wage and salary earners in the total number of employees in Greece in 1981 amounted to 43% against an average of 90% in the nine EC member states of that time. See K. Vaitsos and T. Yannitsis, op. cit., p. 26. On the problems of employment in Greece see especially the interesting work of C. Tsoukalas. State, society and work in Greece after World War II., ed. Themelio, Athens 1986, p. 199 f.

Thus, Greece has the lowest share of wages and salaries in the GNP of all OECD countries. In 1982, it was 39%. In contrast to this, in Germany for instance it was 56%, and in the USA 61%, Ibid., p. 243.

The problems of Greek industry are placed on the micro rather than on the macro level. See T. Yannitsis, op. cit., p.361 f. They are mainly connected with:

a) the preponderance of traditional production branches, the importance of which has increased considerably in the last years,

b) the low productivity which, among other things, is to be attributed to the predominance of small and medium sized enterprises, and, along with this, to the low concentration degree and,

c) the missing inter-sectoral specialisation, and last but not least with d) the labour intensive character of the production methods used.

As it has been pointed out, the main weaknesses of Greek industry are the technologically and organisationally obsolete structures of production and the absence of modern efficient management. At all times Greek management has been guided by traditional entrepreneurial patterns of behaviour and often opposes new methods of business management and labour organisation as well as technological innovations as long as they are not absolutely necessary.

As far as business organisation and administration are concerned, the traditional ideas based on the personal dominance of the proprietors still prevail. Trapped in their "ownership prejudices", Greek entrepreneurs usually employ irrational methods of business administration. Afraid of losing control (a very great percentage of Greek enterprises are family-owned) they refuse to delegate responsibilities to subordinates and thus prevent them from developing their capabilities that could otherwise contribute decisively to improvements in the operation of the enterprises and therefore increases in their productivity. See Z. Papadimitriou, Prospects for Anthropocentric Production Systems in Greece, APS Research Papers Series, vol. 17, Commission of the European Communities, July 1991, Brussels, p. 20.

The Greek economy was not directly affected by the global Fordist model of economic development due to the peculiarities in its structure (i.e. the large number of small units, low degree of verticalisation in production etc.) and also to the fact that it was integrated in the international division of labour only after a certain delay. An exception is the heavy industry which could be characterised as "peripheral fordism" according to the definition given by Alain Lipietz, comparing the productive systems in the southern European countries (Greece, Portugal and Spain) with those of the metropolitan centres of Fordism. In other words, the Fordist model of the organisation of production has never been systematically applied in Greece. Though in some cases certain forms of "primitive Taylorism" (Lipietz) are applied, the organisation of labour as well as the organisation of production, particularly in the small and medium sized enterprises, is generally implemented in an unsystematic and empirical fashion. Referring to the model of "peripheral Fordism" and the countries of the European South, Greece, Portugal and Spain, see A. Lipietz, Mirages and Miracles - The crisis of global Fordism, Verso, London 1987, p. 113 f. In the case of Greece, see also A. Lyberaki, op. cit., p. 140 and Th. Pelagidis, Southern Europe in the era of post-fordism : The case of Greece, M. Phil. Thesis at the University of Sussex (IDS), 1989, p. 22 f.

As to innovative behaviour, a large part of Greek entrepreneurs in the manufacturing industry are guided by the logic of fast and easy profit (short-term payback criteria). Often they are not prepared to invest in new technologies and therefore the technological standing of Greek industry is relatively low. There is an urgent need for technological and organisational modernisation of Greek industry if it is to withstand the growing pressures of competition.

The model of so called "flexible specialisation" could be a possible way to overcome the production and marketing problems of the Greek economy. Flexible specialisation concerns the way products are produced as well as the products themselves. It is achieved:

a) by promoting small-scale production in technologically advanced areas, and

b) through decentralisation within the enterprises.

The model of flexible industrial production is based on the potential of Flexibility of the new technologies, and on the use of qualified labour. It is a reaction to the exhaustion of the possibilities of productivity increases in the model of full automatization. Main characteristics of flexible industrial production are small-scale production, product differentiation, implementation of machines with multiple usage possibilities. Regarding the problematics of flexible specialisation, see M. Piore and Ch. Sabel, The second industrial devide - Possibilities for prosperity, New York 1984, Scott, "Flexible Accumulation and regional development: The rise of new industrial spaces in North America and Western Europe, in: International Journal of Urban and Regional Research, 12 (1988), pp. 171-186, M. Storper, "The transition to flexible specialisation in industry", in: Cambridge Journal of Economics, 9 (1988) as well as A. Lyberaki, op. cit., p. 17 f. and Th. Pelagidis, "Flexible specialisation as a development strategy", Nea Kinoniologia (New Sociology), 10(1990/91), p. 47 f. (in Greek).

Even though there are many, and to a great extent, well justified doubts on whether and to what extent flexible, specialisation forms an alternative model of economic development, appropriate to provide answers to the problem of diminishing productivity and falling profits. See A. Sayer, "Post-Fordism in Question", International Journal of Urban and regional Research, 4 (1990). No one can deny its advantages, particularly in those aspects concerning the development prospects of the small and medium sized enterprises. In this regard, flexible specialisation becomes a particularly attractive strategy for Greek Industry and manufacturing in general.

The technological infrastructure of Greek industry

The post-war economic development in Greece was based almost exclusively on the transfer of technology from abroad. The technological standing of Greece today is characterised by its dependence on industrial production of investment goods, the dependence of the technologically advanced branches on foreign capital in the industrial sector, and finally by the country's inability even to produce technological know-how necessary for the development of the traditional industrial sectors.See T. Yannitsis, op. cit., p. 361 (in Greek). On the technological development of Greece see more details in S. Skoumal and D.A. Kazis, op. cit., p. 42 f. as well as N. Komninos and others, Strategical priorities for the technological development of Greek Industry - Social, sectoral and spatial consequences (technolocations and industrial innovation), Final Report, ed. Aristotelean University od Thessaloniki 1987, p. 51 f. (in Greek).

The degree of diffusion of new technologies in Greek industry varies from branch to branch. While large enterprises, particularly those affiliated in innovative ways with foreign capital, are relatively open to new technologies, the innovative readiness of the small and medium sized enterprises, is very limited. Besides the restraining influence of the low degree of production verticalisation and the fact that production units are small, technological modernisation of enterprises is handicapped by the attitudes of enterpreneurs themselves. Usually, they prefer to purchase complete technical solutions abroad, instead of contributing to the development of technology domestically. When they decide to import technology, they use short-term criteria, which do not permit the potential of new technology to be fully utilised. Generally, Greek enterpreneurs show a basically instrumental attitude towards technology, which prevents the development of a long-term strategy for the economy's technological modernisation.

With the exception of the banking system and insurance companies, the degree of diffusion of new technologies into Greek economy and particularly in the industrial sector is limited. The new technologies introduced in industry are used mainly for administrative purposes while relatively conventional technology is used in the directly productive process. Only in recent years has there been a tendency to adopt modern technological systems such as robotics, Computer Numerical Controlled Machine Tools (CNC - Technologies), Computer Aided Design (CAD) and Computer Aided Manufacturing (CAM) etc. The use of Computer Integrated Manufacturing (CIM) and of Flexible Manufacturing Systems (FMS) is almost non-existent.

Recently, serious efforts are being made to strengthen the enterprises' readiness for innovation. At this moment, three Business Innovation Centres (BIC) are operating in Greece. These offer significant services to Greek enterprises on issues relating to technology transfer, executive training, development of new technology, know-how and innovations, new product design, organisation of production with the assistance of new technologies, etc.). See Z. Papadimitriou, op. cit., p. 19 f.

But this is not enough. Compared to other member states of the EC, Greece significantly lags behind in scientific and technological research. This holds particularly true for industrial research. On the state of industrial research in Greece, see Ministry of Research and Technology, Industrial Research: Proceedings of an assembly for the promotion of co-operation between the productive and the research sector, Athens 1984 (in Greek) as well as L.Tsipouri, op. cit., p. 84 f.

Greece spent only 0.33% of the GDP for research purposes in 1986, compared to 2.70% in USA, 2.74% in Germany and 2.78% in Japan. The state contributes 74.4% of the funding for scientific and technological research, enterprises contribute 23.2% while 2.8% of research expenses are provided by other sources (the data for Japan for instance appears as: state 21.3% enterprises 68.7% and others 10.0%). See Ministry of Industry, Energy and Technology (General Secretariat of Research and technology), Science and Technology-Indicators of Scientific and Technological Research, Athens 1989, p. 17 (in Greek).

Problems related to Research and Development (R & D) in Greece are not only of a financial, but also of an organisational nature. The lack of co-operation between research institutions and the productive sectors of the economy is characteristic. This absence of co-operation impedes the proper disposal of the research results, which often remain unexploited, at the expense of the development of the Greek economy in the long run. See Z. Papasimitriou, op. cit., p. 25.

Shortcomings in scientific and technological research generally, and particularly in industrial research, are due to a great extent to the development policies implemented by the post-war governments in Greece. During that period, the dominant belief was that investing in the area of R & D does not pay economically in a small country like Greece. Thus, instead of promoting research within the country, a policy was pursued by which the economy became completely dependent on the profitable usage of other countries' technological achievements .

4 Social aspects of the adjustment process

The common EC domestic market will fail to achieve its social aims if there is no aspiration for a similar harmonization of the working and living conditions of the employees within the EC. It has been rightly pointed out that the project of a common domestic market with its driving forces, motives, and ambitions is much more then merely an economic affair. It means the attempt to develop Europe to be an independent and globally acting political-economical power See O. Jacobi, "Gemeinsamer Markt 1992 - Wirtschaftliche und politische Hoffnung?" (The common home Market 1992 - A hope for economic and political development?), Paper given at the expert meeting of the OETV Hessen, 25.10.1989, p. 3.
As it is well known, the differences concerning working and living conditions between the EC countries are extensive. This, by the way, seems to be the reason why the Commission of the EC was motivated to pass the "Charter of social Basic Rights"See Commission of the European Communities, "EC Charter of Social Basic Rights", com (89) 471, Final Draft, Brussels, 2.10.89. In this respect, too, Greece has to overcome as series of difficulties which are mainly to be attributed to the not particularly well developed condition of industrial relations. Because of historical, political, economical etc. reasons which cannot be dealt with here in detail, the Greek workers' movement did not succeed in getting up with the western European union movement. Therefore industrial relations have become stuck on a relatively low level.The main characteristics of the Greek industrial relations are:

a) close connection between unions and political parties, so that the target of trade union organisations are often identical with those of the political parties with which they are affiliated,

b) intense government intervention stemming from each government's efforts to ensure union support in order to implement it's economic policy,

c) splitting of the trade unions into numerous first and second degree organisations, which creates functional problems in the relations between them,

d) the employers' lack of interest in dealing with conflicts with the labour force through consensus procedures, and

e) the strong presence of clientele relations and of corporatist thinking in the trade union movement, particularly in the public. See Z. Papadimitriou, op. cit., p. 22.

A characteristic feature of industrial relations in Greece is the focus of collective bargaining on purely economic and institutional demands, as well as the lack of a systematic treatment by the social partners of problems concerning employees working conditions. The discussion regarding the upgrading of work and the improvement in employees' working conditions has just taken place. The fact that working conditions have not been the object of bargaining policy in Greece for a long time is due mainly to:

a) the absence of any industrial tradition,

b) the absence of a substantial interchange of thoughts among the social partners about the future of human work, and

c) the weaknesses of the Greek research system, especially the low level of socio-economic research, Ibid., p. 2.

In order to overcome these difficulties concerning the industrial relations being able to participate in the new developments in a united Europe for the improvement of the working and living conditions of the Greek employees, presupposes a radical re-orientation in the behaviour of the social partners (trade unions and employers). Whereas the trade unions, but also the employees themselves, will have to overcome their instrumental attitude towards work and their purely economic demands, the employers are called upon to renounce their short-term, purely profit-oriented opinion of production and enterpreuneurship. The enterprise is not merely a place of production but also a social unit that must, apart from striving for profits, also satisfy social demands, Ibid., p. 26.

5 Conclusions

At present, the Greek economy finds itself in a far-reaching crisis.

Tendencies of de-industrialisation, accompanied by an immense indebtedness of the public sector, indicate that the reorganisation of the economy is an absolute necessity if the country is to meet the requirements of the common home market. In the face of diminishing competitivity of Greek products (low productivity, missing quality standards etc.) an acute question poses itself for a new strategy of industrial development which could lead to an overcoming of the structural problems of the industrial sector. It is true that at this moment any experience in dealing with alternative forms of production is lacking, but with closer inspection of the Greek economy it can be stated that certain structural characteristics exist which could become the basis for the project of a model for production and development oriented at flexible specialisation. There are to be named above all:

a) the large number of small and medium sized enterprises, and

b) the still wide-spread craft tradition in some areas of manufacturing industry and the respective existence of qualified labour.

Bearing in mind the structural characteristics of the Greek economy, " flexible specialisation" could be not only a solution to organisational problems in production and marketing and also the ability for adaptation to demand fluctuations, but also an improvement of working conditions. The flexible industrial specialization is an attractive strategy, especially suited non-Fordist economies which, for their competitiveness, rely mainly on their capacity for quick response to changes that take place within the dominant countries of the world economy.

The realisation of a new model of production presupposes a national technology policy that goes far beyond the mere transfer of techniques, machines and technical know-how, creating a framework favourable to the development of a technology that takes account of the economic, social, cultural, and ecological conditions of the country. In order to achieve this, an elaborated programme of Research and Development (R & D) has to be installed, equipped with, naturally, the necessary financial resources. Research and Development must not be left to the state alone. There is a need of stronger engagement of the enterprises.

16 Class struggle in Greece and the EEC: Accession to the EC and Greek social interests

Andreas Moshonas

I

This article is actually a summary of a research study carried out by the author from 1987 to 1989 at the Foundation for Mediterranean Studies in Athens.[1] The study deals with the socio-political effects of the Greek Accession to the European Economic Community (EEC), with its effects on the socio-political domestic struggle. In this sense, the study focuses on "political matters", related to the reproduction or transformation of relations of exploitation and power. It is, therefore, an analysis of socio-political struggle processes; mainly in the sense of class and political struggle in general. Thus, an investigation is virtually made on the relation between the Greek accession to the process of European integration on the one hand, and political strategies for maintaining or achieving class domination on the other. Social classes and strata - collectively expressed through specific trade-unions and professional associations as well as political parties and state power are included in this context of contrasting relations, as basic elements of action.

The first volume, which is the study summarized here, examines the formation of social interests in the context of the Greek accession to the EEC. The study has seven main chapters, two introductory ones, a conclusion and bibliography.

In the first two chapters, we present the research object from both a general (Chapter 1) and a specific (Chapter 2) aspect, outlining the theoretical framework of the analysis as well as the order of presentation.

The next four chapters deal with the formation of interests by the various categories of capital: industrial (Chapter 3), commercial (Chapter 4), banking (Chapter 5) and transport (Chapter 6).

Chapters 7 and 8 analyse the formation of interests concerning the petty bourgeoisie and the rural strata, respectively. Chapter 9 examines the interests of the working class, and Chapter 10 presents some general conclusions. Further on, a summary of all these conclusions is given in the context of a theoretical generalization.

II

The Greek Industrial Capital

The Greek industrial capital has been analysed in the context of its accession to the process of European integration. This process is characterised not only by the liberalization of the economic and commercial transactions, but also by the absence of a Community industrial policy, capable of protecting the less developed countries. The function of the industrial capital is also conditioned by national objectives which according to the respective ideological and political orientation of governments are extended from the simple modernization of industrial enterprises to the restructuring of industry itself. The industrial bourgeoisie having unconditionally accepted the Accession and being powerless to upgrade its own role through the restructuring of production has led itself to defensive tactics, mainly aiming at modernizing the existing structure, so as to be simply reproduced. This resulted into their taking flight towards state protection, under conditions of intense competition.

The Greek Commercial Capital, analysed within the framework of Community liberalisation regulations as well as respective national issues and pursuits, focuses on two points: a) the challenge of imports, due to the creation of the Integrated Community Market and b) the imperative need for extroversion, mainly due to the limited extension of domestic demand. In this tug-of-war, the representatives of the Greek Commercial capital aim at obtaining the greatest possible liberalization of imports and state support for their exporting efforts.[2]

The Creek Banking Capital is closely connected with the process of the internationalization of relations of production and of exchange. In the context of European integration, this trend has been crystallised into a tendency towards liberalizing the money and capital market. This liberalization affects not only the status of the inflow-outflow of capital, but also the installation of foreign banks in Greece. As a consequence, competition in the money and capital market becomes keener. This results into reinforcing the tendency towards not only a free adjustment of the bank interest rate, but also an increase of labour productivity in the banks, due to the use of modern processes during the performance of their activities.[3]

The Creek Capital of Commercial Transports, analysed within the framework of the relevant Community regulations, includes two categories:

- Firstly, the ship owner´s capital, the representatives of which aim at ensuring the maintenance of its historically internationalized role - in a

Community regime of controlled liberalization - in order to face the negative consequences, deriving from the increasing international competition in freight transport.

- Secondly, the capital invested in the enterprises of road freight transport, which are mainly enterprises of low capital accumulation. This capital aims at ensuring state support and protection, in order to face the consequences deriving from both the Accession and Community competition.[4]

The Greek petty bourgeoisie has been examined in the context of Community regulations, which mainly aim at incorporating small and medium enterprises into the network of operation of big capital, and the respective national concerns. The acceptance of the Accession and the liberalization resulting from it, tend to make state protection a necessity, while the Community regulations for small and medium enterprises tend to legalize, to some extent, national interventions, which aim at supporting and developing these enterprises.[5]

The Greek Rural Strata have been analyzed within the framework of the particularities of the agricultural economy and the Community policies for agriculture. The above mentioned strata are dominated by self-employed farmers, who constitute the focal point of the formation of interests and present a high degree of sensitivity towards matters of: investments, cost of banking money prices of products and income subsides. In a process of reforming the Community policies for agriculture and of restrictive price-policies, the representatives of these strata tend, not only to support the basic principles of the Common Agricultural Policy but also to maintain or even extend state protection. [6]

The Greek Working Class has been analyzed with reference to both the relevant social and economic aspects of Community regulations and the respective national objectives: namely, promotion of growth and capital accumulation, in association with the restrictions imposed by the labour market and the need employment increase. It is in this context that the need to modernize the production process becomes imperative, bringing questions of: labour cost, working conditions and functioning of labour institutions into the foreground. They arc issues which historically constitute the focal point of class struggle. [7]

III

Class and political struggle, in its widest sense, results from specific relations of production and reflects a network of contradictions and antagonisms, developed in the context of contemporary Greek society. On the one hand, there is the main contradiction of the capitalist mode production, namely the contradiction of the capitalist mode of production, namely the contradiction between the social character of production and the individual appropriation of its result, formed as a contradiction between two classes: the capitalist class and the working class. On the other hand, there are secondary contradictions traversing the wider social formation. They have as a focal point the contradictions between the various categories of capital, between capital and petty bourgeoisie/rural strata, as well as between the working class and the petty bourgeoisie. More specifically (a) concerning the contradiction between the capitalist class and the working class, Greek capital faces - in the context of the Accession - the particularities of its historical formation and the resulting negative consequences of European integration. The historical dependence of Greek capital on international capital has forced the Greek capitalist class into accepting the Accession and the resulting economic liberalization, as a suitable means of economic survival and development. But at the same time, the liberalization of the economic and commercial transactions - within the framework of the Integrated Community Market - entails a keener competition inside the Greek economy, thereby making necessary the modernization of the production process and the appeal for protection In this way, the Accession becomes, for the capitalist class, a means to legitimise its specific interests, which undergo the pressures of both the process of capital accumulation and the consequences of European integration.

Protecting the interests of the capitalist class the substance of which is examined further on is entering into keen competition with the working class, due to their historically conflicting interests. In this context, the concerns raised by the representatives of capital concerning the contradictions on the accumulation process, and the effects of the Accession - are interdependent with the corresponding concerns stated by the representatives of the working class.

First of all, the duration of the working day as a factor determining the rate of exploitation of the working class is a matter of immediate interest to both capital and labour. Therefore, the need to gradually harmonize the Greek practices with the corresponding Community regulations, in combination with the initial acceptance of the Accession by the representatives of capital. had as a consequence the weakening of the reactions of the capitalist class and the political legitimization of the claims set up by the working class. So, the Accession to the Community itself and the consequent acceptance of its regime tended - despite the substantial absence of appeal to the respective Community regulations by the representatives of the working class—to function as a factor supporting the position of labour in its struggle against capital, for reducing the working hours. This Community legitimization made in its turn, easier the already required state regulation on the critical question of the duration of the working day; namely of the reduction of the working hours.

While the duration of the working day sets the limits of the absolute surplus value appropriated by capital, the modernisation of the production process and the resultant increase of labour productivity determine the extent of the relative surplus value. Therefore, the increase of labour productivity decreases production costs, consequently augmenting the surplus value rate appropriated by capital. The Accession to the Community and the liberalization ol transactions, within the framework of the Integrated Community Market, aggravates competition, thus resulting in an imperative need for modernizing the production process, mainly through an increase in the technical composition of capital. On the other hand, the particularities in the formation of Greek capital, associated with its general reproduction crisis which actually tends to become more acute in the context of the Accession creates refolding tendencies in the Greek capitalist class; namely trends of maintaining and reproducing the existing productive structure. The result of these two processes: i.e., "modernising" and/or "maintaining" the productive structure, is an aggravation of the struggle between the capitalist class and the working class, concerning two issues of great importance:

- Firstly, the tendency to maintain and/or reproduce the existing productive structure which characterises anyway the greatest part of the Greek capitalist class - creates an imperative need to augment the surplus value by other means. namely by:
  a) increasing the intensiveness of work, which is actually a covered extension of the working day, where the employed provides more work in a 'condensed- way and,
  b) restraining wages. Therefore, it is not accidental that within the framework of the Accession's consequences the struggle of the working class has focused on issues of working conditions and mainly of wages. The latter was found to be particularly under conditions of inflationary pressures. This mainly occurred as a result of capital's efforts to ensure important profits and/or the simple survival of business activity in the country, by maintaining prices and products at relatively high levels.

- Secondly, the tendency to modernize the existing productive structure, especially when it succeeds through an increase of the organic composition of capital, entails a reduction of employment. Therefore, it is not accidental that the modernization effort made necessary within the framework of the barriers to the accumulation process and the Accession effects - is accompanied by the intensified struggle of the working class, which aims at maintaining employment and improving working conditions.

The above mentioned conditions include issues of: equal treatment between men and women, mass lay-offs, health and safety, and generally regulation of industrial relations and working time. Therefore, they concern matters which affect the rate of exploitation of the working class. however, the Community has adopted, in the context of the Community contradictions and the class struggles, labour law regulations providing for the participation of working people in labour institutions of consultation and deliberation. So, inspite of the Greek labour representatives' substantial absence of appeal to these Community regulations, the acceptance of the Community regime by the Greek capitalist class tends to reinforce the political legitimization of demands and the struggle of the working people concerning industrial relations. But, in the context of the process for the development of an integrated social area in the Community, this political legitimization, derived from the Community, tends to be reinforced. This occurs to the extent that the trade-union movement of Greece is in a position to: a) develop a militant strategy with European orientations and b) exploit the contradictions of capital itself, being necessarily aggravated by the consequences of the Accession and European integration.

(b) Inter-class contradictions in the area of capital appear in the first place in the production process among the categories of productive capital. On the one hand, there are categories of capital which, through a modernization process of production, manage to maintain production costs on relatively low levels, increasing therefore their profit rates, under conditions of growing competition. On the other hand, there are categories of capital, which function under conditions of relatively high production costs. This makes their extended reproduction, either difficult or even impossible, without state support. The Greek Accession to the process of European integration tends to aggravate these contradictions, thereby reinforcing the regulatory role of the State and increasing, at the same time, the intervention possibilities of the trade-union movement in the very process of development and of capital accumulation. And this with the aim of promoting the modernization process of the economy, not necessarily oriented towards its adaptation to the Community regime (capitalist integration) but mainly towards the transformation of the mode of accumulation itself.

During the production and realization process of surplus value, productive capital functions along with commercial capital. On that point, while the function of the letter is supplementing that of the former their interests are conflicting. It does occur since the surplus value, appropriated by commercial capital, constitutes a part of the surplus value, produced by the main production process itself. Within this framework, the Accession of Greece to European integration tends to aggravate the contradiction between productive and commercial capital to the extent that, due to the existing competition, it restricts the rate of surplus value and thus makes imperative the modernization of the production process. This modernization tendency appears clearly in two ways. On the one hand, productive capital undertakes certain functions of commercial capital, in order to increase the surplus value which it appropriates. On the other hand, commercial capital becomes a pole of attraction for productive capital. This tendency takes place to the extent that the conditions of operation of the Integrated Community Market increase in Greece the profit margins of commercial capital, comparatively to those of productive capital.

The latter, within the framework of economic liberalization and the resulting challenge of imports, has to confront the (importing) commercial capital as far as the volume and the prices of the imported goods are concerned. But, due to the Accession-s acceptance by the capitalist class, this contradiction tends to fall within the competence of the State. It clearly appears at the issue of import's liberalisation in two cases: a) concerning the restriction of imports, in products creating an acute competition to the respective domestic production, and b) concerning the free admission of imports, in products (capital goods and raw materials) which constitute the inputs of the production process. In the process, however, towards the completion of the single European market, the resultant challenge of imports makes state support an imperative need in both exports (within the framework of the endemic extroversion of the Greek economy) and the domestic production process.

These contradictions pass through the whole circuit of capital, creating a confrontation between productive-commercial capital and money-capital, in the context of money and capital markets. The Accession of Greece to the process of European integration leads to the liberalization of money and capital markets. This results in strengthening the relative autonomy of banking capital, in the context of the extended reproduction process. It means that the Accession tends to restrict the State's regulatory role, concerning the value of banking money. It results in creating margins for increasing the rate of surplus value, appropriated by banking capital; that is increasing bank profits, through the respective adjustment of the interest rates. Therefore, the liberalization of money and capital markets tends to aggravate contradictions among the various categories of capital and to lead - through the modernization process of the whole circuit of capital to a deeper fusion of the functions performed by productive, commercial and money capital.

The capital of freight transport, necessarily involved in this process as well, constitutes an important factor during the realization of surplus value. The liberalization of economic and commercial transactions, entailed by the Accession to the Community, tends to aggravate contradictions with the capital of transports, mainly as follows:

- Firstly, the liberalization of money and capital markets and consequently the exemption of bank interest rates from restrictions, tend to aggravate contradictions between the capital of transports and banking capital, concerning the cost of money? namely the level of the bank interest rate. In the context then of the particularities of the Greek economy, this contradiction refers more to the capital of road freight transports directly dependent on the Greek banking system than to the ship owner's capital, internationalized since the time of its historical formation .

- Secondly, the liberalization of freight transport - within the framework of the Integrated Community Market - liberates transport freight from the existing administrative regulations. This results in aggravating contradictions between productive and commercial capital to the extend that the conditions of operation of the transport market lead to relatively high freights.

- Finally, the liberalization of commercial transactions and the establishment of an integrated European market are processes forcing productive capital to undertake a part the function concerning not only commercial capital, as indicated above, but also the capital of transports, thus aggravating contradictions between industrialists and transport agents.

(c) The Accession of Greece to the process of European Integration contributes to: a) the deeper incorporation of the agricultural economy into the industrial and urban economy and b) the aggravation of contradictions between the capitalist class and the rural strata. To begin with, in the context of the Common Agricultural Policy of the Community, prices of agricultural products have to be adjusted on the basis of Community rather than national criteria, thus resulting in the presence of social friction mainly towards two directions:

- Firstly, towards the industrial and commercial capital. This occurs to the extent that agricultural prices (and consequently rural incomes) present all important deviation from the prices of industrial products (mainly from those used as means of production in the agricultural economy). The latter prices level derives from the monopoly conditions of the market.

- Secondly, towards the State. This occurs to the extent that prices of agricultural products are not in position to cover production costs and /or to secure a satisfactory rural income.

The contradiction between the rural strata and the capitalist class is presented not only in the market concerning prices, but also in the production process itself; to the extent that capital penetrates into the agricultural economy, thus creating capitalistic agricultural holdings and enterprises. Actually, this process of capitalistic evolution concerning agricultural production is strengthened by a) the Accession of Greece to the Community and b) the resultant deeper integration of the agricultural economy into the functions of capital, both at the national and the international level. In this context. the contradiction between the capitalist class and the rural strata takes a double form: a) in production, as a contradiction between capital and labour, due to the creation and reproduction of a rural proletariat for the functioning of agricultural capitalistic enterprises, and b) in the market, as a contradiction among the producers of goods, due to the aggravation of competition between the production of capitalistic enterprises (for example, production of green houses) and the respective production of family agricultural holdings .

The aggravation of competition in agricultural production, in combination with the shrinkage of margins related to price increases of agricultural products - in the context of the restrictions imposed by the Community agricultural policy - tend to increase the sensitivity of the rural strata, not only towards the matter of structural changes, aspiring to productivity growth and improvement of competitiveness, but also towards the redistribution of social surplus. The claim for the latter, which necessarily leads to conflicts with the State and the capitalist class, focuses on two issues:

- Firstly, on money *inflows* to the agricultural economy (investments and fiscal subsidies) with emphasis not only on the magnitude of these inflows but also on the cost of money (banking interest rate), which tends to increase within the framework of liberalizing the money and capital markets.

- Secondly, on the money outflows from the agricultural economy, with emphasis not only on the prices of the agricultural products and on the terms of exchange but also on the taxation of the agricultural production and consequently of the rural incomes.

(d) The redistribution of social surplus constitutes the focal point of the concerns raised by the petty bourgeoisie, in the context of both the particularities of the Greek production process and the Accession's effects as well. In particular, the Accession of Greece to the Community - in association with the entailing aggravation of contradictions in the area of the small and medium enterprises becomes an "alibi"to the petty bourgeoisie in order to legitimize its positions and promote its claims to the State. It is in the context of the social surplus redistribution that serious concerns are raised, (so as to secure - as a means of business survival and development - special taxation treatment and significant bank financing. As far as this latter is concerned, the interest focuses not only on the availability of bank money but also on the cost (interest rate) of the bank loans. This factor tends to be particularly encumbered in the course of time, due to the compulsory liberalization of the banking system, within the framework of establishing the Integrated Community Market. Therefore, in a regime of free formation of interest rates by the banking system, conflicting interests tend to be developed between the small and medium enterprises and banking capital.

Additionally, the market's liberalization, in the context of Community regulations, tends to aggravate the contradictions between the small-medium entrepreneurs and the capitalist entrepreneurs on the cost of the means of production (raw materials and capital goods) used by the former enterprises.

This happens, not necessarily because of price increases in the means of production supplied by capitalistic entrepreneurs (industrial-commercial) to small and medium enterprises, but also due to the compression of the profit margins of the latter enterprises in the regime of the Integrated Community Market. Therefore, price control over production inputs constitutes here as in the case of agricultural production - a necessary "escape" from the structural problems of the small and medium enterprises themselves.

Increases of labour costs (wages and duration of the working day) which take place compulsory in the context of the harmonization with the Community regime, function in the same course of action. However, compared to the position of big capital, small and medium entrepreneurs (mainly those politically oriented towards progressive course of action) tend to form their positions in a framework of contradictory objectives. On the one hand, the regulation of labour costs creates favourable conditions for business survival and development. On the other hand, the regulation ,labour costs creates obstacles to the effort of hammering the alliance between the petty bourgeoisie and the working class. Thus, the Greek Accession to the Community seems to function against this alliance, to the extent that small and medium entrepreneurs are not able to absorb socially acceptable labour cost mainly through the rationalization of their own business activity.

IV

It is worth noting here that the modernization effort, concerning the production process, is strengthened within the framework of the Community regime not only because of the pressures imposed on the production process by the function of the Integrated Community Market, but mainly because of the special measures of Community policy aiming at the modernization of small and medium enterprises and the agricultural economy as well. But, the application of these modernization measures, in association with the market pressures brought in a regime of transactions' liberalization, tends to create a differentiation to interests inside the petty bourgeoisie and the rural strata to the extent that it creates and reproduces competitive and non-competitive enterprises. This means that the Accession to the Community tends to impel - through the modernization process resulting from it - the creation of contradictory situations: that is the formation ol converging or conflicting interests and the moderation or aggravation of contradictions all over the social structure, including the area of the petty bourgeoisie and the rural strata.

This research has nevertheless showed that the above mentioned differentations have not yet been crystallised at the level of forming social (class) interests, since the modernization process itself of the Greek economy and society is still at its initial stages. So, it is not accidental that the formation of social interests - realized in the "tug of war" created by the contradictions and particularities of the social structure and the political practices in Greece usually follows traditional processes based mainly on clientelistic relations.[8] But how long will this continue? It is not easy to answer! Nevertheless, there is one indisputable fact: the Greek society changes, and this is certainly occurring at a quick pace.

Notes

1. Moschonas, Andreas (1990), 'Class Struggle in Greece and the EEC', Vol. 1: *Accession to the EEC and Greek Social Interests* (in Greek), Foundation for Mediterranean Studies, Athens.
2. c.f. Giannitsis T (1988), Accession to the EEC and Effects on Industry and Trade (in Greek), Foundation for Mediterranean Studies, Athens.
3. c.f. Seremetis D (1989) Financial Relations between Greece and the EEC (in Greek), Foundation for Mediterranean Studies, Athens.
4. c.f. Tsetsis Stavros.: The Common Transport Policy of the EEC (in Greek), Evrokoinotikes Ekdoseis, Athina; no date.
5. c.f. Moschonas Andreas (1986) The Petty Bourgeoisie: the Case of Greece (in Greek), Foundation for Mediterranean Studies, Athens, for an analysis of the formation of the Greek petty bourgeousie.
6. c.f. Maravegias Napoleon (1989) The Accession of Greece to the European Community: The Effects on the Agricultural Sector (in Greek), Foundation for Mediterranean Studies, Athens.
7. c.f. Tzekinis Christos (1987) The Greek Labour Movement, 1870-1987 (in Greek), Galaios, Athina, for an analysis of the formation of the Greek working class.
8. c.f. Mouzelis Nicos (1978) Modern Greece: Facets of Underdevelopment, Macmillan, London, for a historical analysis of these questions.

Transformation of Patterns IV
CONCEPTIONS OF MODERN TECHNOLOGY

17 The challenge of technological progress in modern societies

Detlef Jahn

Introduction

Social and technological changes have been tremendous in recent years. Industrial society is entering a new stage of development. After the severe damage to the vision of socialism through the ideological breakdown of Eastern Europe, Western capitalism is presently increasingly in question. There are some indications that the cultural core of industrial society is being challenged, which may lead to disintegration and new or re-vitalised tensions and conflicts. This new situation may even separate otherwise coherent social units.

The cultural changes in highly industrialised countries are manifold. In this article I wish to focus on one specific problem. The problem that emerges from the challenge of technological progress. This aspect will be analysed in both theoretical and empirical terms. The empirical analysis is essential since theoretical conclusions take rather different empirical forms among Western societies.

Technological progress is a major concept for the understanding of the development of industrial society (Alexander and Sztompka, 1990). However, technological progress is not a value-free concept. One may ask who and what influence technological, progress and what function has it in the social and political organisation of society. In this paper I will focus on the function of technological progress serving as a means of societal consensus. Due to the fact that in modern societies organisations create, stabilise and change social processes (Perrow, 1986), organisations are an important part of this analysis. The significant role of organisations in the process of shaping technological progress (Rammert, 1992) and societal consensus (Jahn, 1992c) has been stressed elsewhere. However, the role and function of technological progress have been challenged in recent years. The most obvious examples are new social movements which undermine the societal consensus that builds upon technological progress and economic growth (Eder, 1990a).

It is a crucial function of technological progress to promote a compromise between labour and capital. In this paper I want to address the question whether or not this compromise has been challenged by the criticism of technological progress. In order to investigate this question empirically, I analysed the capital's and labour's standpoints towards technological progress. Putting it in more concrete terms I focused on the expressions of employer associations and trade unions in Sweden and Germany.

By considering Sweden and Germany, I wish to indicate substantial differences in the challenges confronting technological progress in both national settings. Both countries are particularly interesting in this respect. They both belong to the most advanced industrialised countries in Europe; however, they may differ in their treatment of technological progress. In the field of environmental policy, for instance, Sweden belongs to the leading nations when it comes to this issue. Already in the late 1960s and early 1970s, Swedish policy-makers quickly responded to the environmental challenge. Concerning energy policy, Sweden has been seen as a model for restructuring theirs. Therefore it is no wonder that Sweden has been taken as a model for a country that may favour a social development that is less committed to the productionist line and more open to new demands which place less emphasis on technological progress and economic growth (Lindberg, 1977; Kitschelt, 1983; 1986; Sahr, 1985; Gottweis, 1986). Furthermore, Sweden is also of interest, because the Swedish labour movement has been able to shape Swedish society. Germany, on the other hand, is of special interest here because of its strength in new social movements and also the Green Party which has led the eco-ideological debate for more than one decade (see, e.g. Brand (ed.), 1985; Klandermans (ed.), 1989; Müller-Rommel (ed.), 1989; Dalton and Kuechler (eds.) 1990; Jahn (ed.) 1993). In addition to which, Germany is also a prime example for analysing the impacts of the change in Eastern Europe, as it is directly effected by it due to the unification of West and East Germany. However, the data used in this paper was only collected until the end of 1990 and may, therefore, not fully grasp this development.

In the analysis, I focused on the nuclear energy debate in both countries. This restriction was necessary for empirical reasons. Before coming to the empirical analysis I wish to outline the form and function of the belief in technological progress and its challenge in recent years. Following some analytical and methodological remarks, I will present the disseminated world views of employer associations and trade unions on technological progress and economic growth over the last two decades. We will find some surprising findings concerning the progressive role of Sweden and the unity of German trade unions.

Theoretical implications

The belief in technological progress

The conventional view on the development of industrial society is rather deterministic and there is a logic present in industrialisation which generates a series of imperatives leading all industrial societies to similar social institutions and practises. The non-questioned premises consist of the following chain of arguments: technological innovations lead to technological progress and to social progress. Social progress is perceived as leading to a fairer, richer and better society. However, progress was initially a religious idea. The notion of linear development towards a utopian endpoint is based in the Judeo-Christian tradition. The reformation brought this heavenly concept down to earth and the enlightenment gave it a fully secular form (Alexander, 1990). The roots of the belief in progress dates back to even the cultural tradition and concept of nature in Greek history (Eder, 1990b) but reached a new quality around the 16th and 17th centuries when the progress of natural sciences had a tremendous impact on social sciences. In particular the positive thinkers such as Auguste Comte, Herbert Spencer, Saint-Simon and Emile Durkheim connected the three elements: technological progress, social progress, better society. Technological progress merged together with the positive assessment of an industrial society (industry in the sense of the quality of hard-working and being usefully employed). Particularly the bourgeois movement of the 18th and 19th centuries transformed these ideologies into political practice. Technological progress and moral and social progress would be just two sides of the same coin. Progress initiated by science and technology, then realised by rationalising the production and labour process would be a means of freeing society from the constraints set by nature and, therefore, would lead to social and moral progress of mankind.

However, the belief that technical progress leads to a better society is inherent in both bourgeois and socialist thinking. The former postulates that free market procedures connected with an unreflecting concept of freedom may develop into a status of morality while the latter believes that technological progress can only function well in a changed society in which technological progress may transform present society to a truly heavenly condition of communism. What is important in these beliefs is that they locate - despite a lot of differences in their interpretation and preference in the organisation of society - the forces shaping society in the process of industrialisation as such. And it is technology which is identified as the engine that drives the system. While bourgeois culture considers social progress as a linear process towards better living conditions, Marxism views technological progress as the promoter of revolutionary change. In this view, technological progress would lead to a development of the forces of production that will finally change the whole social organisation of society, the relations of production. As a consequence, socialist thinking argues, technological progress must be supported even under condition of a capitalist society.

The political implication of the belief in technological progress

The belief in technological progress and economic growth had fundamental consequences in the development of industrial capitalist societies. The concentration on wealth creation made it possible to cover the antagonism between capital and labour. The class conflict has been institutionalised and (temporarily) solved by the class compromise between labour and capital (Dahrendorf, 1959; Touraine, 1971). In this period, consensus politics and economic growth obtained an ideological status (Sabel 1981: 209), which was the unquestioned prerequisite for social welfare. Capital and labour formed an alliance whereby "... capitalism as a growth machine was completed by organised labour as a distribution and security machine." (Offe, 1985: 822)

The institutionalisation of the class conflict has been achieved by the state's withdrawal from obvious class politics, while organisations of labour and capital were gaining an exclusive status. Independent collective bargaining is a prime example in most highly industrialised societies. It should have been their task to autonomously solve the antagonism between capital and labour. The theory of neo-corporatism illuminates aspects of this societal development, although Claus Offe critically focuses on the consequences of public status attributing to interest organisations. I would like to discuss two aspects in this context: first, the interest representation in corporatist societies is conducted by the most powerful interest groups that settle their conflicts at the expense of social categories (groups and issues) which are poorly organised (Offe, 1981: 128/9). In this context, it is important to be aware of two functions belonging to corporate politics:

The advantage of corporatist modes of interest representation over democratic representative ones resides in the potential of the former for depoliticizing conflict, that is, in restricting both the scope of the participants in conflict and the scope of strategies and tactics that are permitted in the pursuit of conflicting interests. (Offe, 1981: 141)

Second, as will be described in more detail below, economic growth obtained a crucial status for de-politicisation. Economic growth served as an ever growing pie from which capital and labour were obtaining ever bigger pieces.

Both sides (capital and labour; D.J.) have an interest in continuous economic growth, one, because it is the precondition for capital accumulation, and the other, because it is a requirement for successful interest representation... both employers and unions have a common interest in the preservation of the capitalist mode of production, the former for obvious reasons and the latter because that mode of production is the basis for their organisational existence, and with its transformation their fate would be uncertain or they would become superfluous altogether. (Erd and Scherrer, 1985: 120)

In this way, economic growth served indirectly other societal goals such as employment, monetary stability, demand stimulation, and even social equity (Schmitter, 1988: 503); both capital and labour could win in this game.

The institutionalisation of the class conflict formed by the consensus on technological progress and economic growth was strongly realised in the two countries under investigation. Of particular interest is the pragmatic belief of the strong Swedish blue collar workers unions confederation (LO) in technological progress that made the compromise between labour and capital possible in Sweden.

In the context of LO's historical perception of technical progress, it is important to note that LO explicitly supports and even promotes it. The quotation from a fifty years old document of LO demonstrates this point:

Rationalisation must be considered a natural, continuing effort to improve the results of production and to enhance the development of human culture. The trade union movement cannot turn itself against these efforts. (LO's 15-mannakommitti, 1941: 144; quoted from Tilton, 1990: 191)

This belief in technological progress was one of the major elements of the co-operation between labour and capital in Sweden (Israel, 1978). However, LO and the Social Democrats, in general, faced difficulties in matching the material goals of this strategy with its resulting consequences. In the 1960s, the costs of technological progress, rationalisation and an implied geographical mobility for the worker created a growing resistance (Martin, 1984; Tilton, 1990: 209-14). Although the elite of LO listened to these complaints and more strongly stressed the qualitative aspects of work-life, it did not abandon its technically-optimistic ideology:

The Swedish union movement demonstrates a unique commitment to rationalisation and efficiency, one that derives from the Social Democratic goal of material abundance, not from an acceptance of capitalist imperatives. (Tilton, 1990: 214)

Labour in Germany has never had the some status and strength as their Swedish counter-part. This is partly due to the fact that the class conflict was not as smoothly institutionalised in Germany as in Sweden and also that German trade unions' ideology has always been more radical than in other countries (Perlman, 1966). However, German unions as well have been known to support technological progress and economic growth like their Swedish counterparts, and this was particularly the case in the post war period when German unions felt obliged to promote the German economy. In that the German unions contributed to the "German economic miracle". Some observers of the German labour movement concluded that labour and capital joined an "economic growth coalition" which is still alive until the present day (Esser, 1982; Alemann, 1983).

Considering this brief presentation of the status and ideology of German and Swedish trade unions, we may conclude that both countries were rather successful in institutionalising the class conflict through a societal consensus that builds on economic growth and an unquestioned belief in technological progress. However, that should have changed in the last two or three decades.

Challenging the belief in technological progress

Until the late 1960s there had not been any substantial doubts concerning the presuppositions of economic growth and technological progress being the driving force for the development of industrial society. It has also been used as a model for less developed countries and regions. Romantic movements of the 19th and early 20th centuries, as opponents to the development of industrial culture, were provoked and suppressed by the dominant culture. The first serious pondering concerning this topic came from experts and scientists who questioned the unbroken continuation of technical progress and economic growth, because they were serious within the code of industrial culture. It became obvious to these people that the technological progress may produce by-products that erode the basis for its further development. The dysfunctional impact of industrialisation on the ecological balance became more apparent. The most out-spoken statements concerning this dilemma have been expressed in the report of the Club of Rome, in the report Global 2000 to the President of the USA, and most recently by the Commission for Development and Environment, the so-called Brundtland Commission. However, the conclusion of these reports mainly focuses on the control of the by-products rather than proposing an alternative concept for social development. The remedies and devices are still integrated in the framework of industrial society.

A more substantial challenge confronting the development of technological progress is being practised by the so-called new social movements. These movements emerged in most of the Western highly industrialised societies in the 1960s and 1970s, questioning the dominant culture and political institutions on a more general base. While struggling for the emancipation of women, peace, and against nuclear power and the destruction of the environment, they have criticised the increasing bureaucratisation and centralisation of modern society. On the one hand, we see that environmental movements concentrate on the harmful effects of industrial development such as pollution and endangering of life due to high-tech facilities. On the other hand, these movements question the basic logic of industrial society, i.e. the belief in technological progress. Therefore, some scholars see a counter-cultural movement that challenges the dominant industrial culture of current society in these new social actors (Eder, 1990a).

One of the most obvious counter-cultural movements is the movement against the use of nuclear energy which challenges the basic values of technological and social progress. Nuclear energy has a symbolic function which was pointed out vividly by Stephen Cotgrove, centralised, technologically complex and hazardous, and reinforcing all those trends in society which environmentalists most fear and dislike - the increasing domination of experts, threatening the freedom of the individual, and reinforcing totalitarian tendencies. Opposition to nuclear power is seen for many as a key issue on which to take a stand against the further advance of an alliance between state power and commercial interests. For the objectors, the material advantages from nuclear power cannot justify the risks involved. (Cotgrove and Duff, 1980: 338; italics added, D.J.)

Undermining the societal consensus

Bert Klandermans (1989) postulates that one of the most important outcomes of the action of new social movements is the successful undermining of the societal consensus. Other impacts of new politics I analyse elsewhere (Jahn, 1992e). By focusing in this paper on the major organisation of capital and labour we can identify whether or not the societal consensus between capital and labour has been challenged through the politicisation of technological progress by the new social movements.

At this point, I cannot go into more depth concerning the actors who have challenged the belief in technological progress, more specifically, the new social movements in Sweden and Germany (Jahn, 1992d). New social movements and their organisations are certainly important variables influencing the impact of the criticism of technological progress and industrialisation on established politics. Here we may conclude that these movements are stronger in Germany than in Sweden and that the initial state response in Sweden has been more open to new demands than in Germany (Kitschelt, 1986; Jamison et al. 1990; Rucht, 1991). However, this fact may have changed over time (Flam, 1992; Jahn, 1992b). In this connection the character and status of the Green parties may have an important impact on the representation of alternative world views. Empirical research also shows that here Germany has a higher potential for "new politics" than Sweden (Jahn, 1993a).

However, in the following paragraphs we will only concentrate on the impact of these factors on labour and capital in both countries. Before doing so, some analytical and methodological remarks are necessary.

Approach and method

For this analysis I used a cognitive approach of collective action (Weick, 1979; Snow et al. 1986; 1988; Gamson, 1988; Eyerman and Jamison, 1991), that focuses on the underlying world views that may determine collective action to a certain degree. The relationship between world views and action is, of course, not perfect; particularly the concept of symbol politics (Sarcinelli, 1987) illustrates this point. However, in the long run social actors may be trapped in the world views which they have presented for years. Robert Axelrod describes the relation between expressed views (assertions) and action as follows:

If a decision maker is heavily constrained to keep his assertions and actions consistent with each other, then he may not only monitor his assertions (to keep them sincere), but he may also monitor his actions (to keep them consistent with his assertions). To the extent that actions are monitored to keep them consistent with the assertions, the assertions themselves constrain action. Because in certain contexts actions and assertions can be mutually constraining, the use of suitable documentary material from which to derive a cognitive map promises to be fruitful in understanding not only the assertion themselves but also the actual choices of the decision maker. (Axelrod, 1976: 253)

This view on social action has its roots in the work of the Chicago School and is associated with names like Charles Horton Cooley, William Isaac Thomas and George Herbert Mead or the concept of symbolic interactionism. Herbert Blumer (1957) and Peter Berger and Thomas Luckmann (1966) transformed elements of this socio-theoretical concept from the micro-sociological level into the research of collective behaviour and societal development. Recently Ron Eyerman and Andrew Jamison have elaborated this concept by combining elements of resource mobilisation theory and the reflection on new social movements by some European authors such as Touraine, Habermas and Melucci. In this analysis social development is an evolutionary process of a cognitive process. At the beginning of a circle of social evolution, new world views are expressed by a small group. They form a new collective identity and help to constitute a social movement. The action of this "new" social movement carves a space for the social discourse in which the new world views are to be expressed, tested and applied (Snow et al.; 1986; Eyerman and Jamison, 1991). Although at first the new world views struggle with the established ones, some aspects of these new ideas are integrated into the established world views.

This concept applied to our research question: the anti-nuclear movements (as a representative of the new social movements) have challenged the world view of industrial society and the basic concept of technological progress has been questioned in the context of criticising nuclear power plants. This criticism started in the early 1970s and has continued to the present day; however the focus of interest lies in capital's and labour's response to this challenge. Before presenting some empirical findings I wish to specify the competing world views of alternative and productionist politics in modern societies.

Alternative and productionist politics: a cognitive approach

The world views concerning the future development of society may be dichotomised into a productionist world view and an alternative one that refers to ecological balance. The former world view can be called productionist since it justifies human action in the light of societal production. The basic value behind this world view is the belief in progress, unlimited growth and improvement. With that, it stands in sharp contrast to world views that stress the importance of sustainability or balance. There is some research done in the field of productionist and alternative world views (Cotgrove, 1982; Kitschelt, 1984; Eyerman and Jamison, 1991). Three dimensions can be distinguished in which productionist and alternative world-views divert most strongly: first, different economic world-views; second, the relationship between social needs and the manner of societal production; third, concepts of societal development and political process. The following brief description of the dimensions should only give an overview about the content of the world views of productionist and alternative development. Since the analysis of nuclear energy is the focal point of my research project, some examples refer to this technology. The statements about the different world views are presented here in an ideal-typical manner, i.e. statements are exaggerated and the connection between them may be less clear in reality:

(1.) Economy: The discussion on alternative and productionist development stresses the fact that there are two competing economic concepts. On the one hand, radical productionists conceptualise costs by means of business accounts. If social costs are included in the business account, they enter it in terms of quantified money terms. Proponents of an alternative development stress to a higher degree "intangible" costs (life, traditions, culture, eco-systems etc.) which cannot be translated into money terms. On the ground of these different preconditions, one can identify conflicting economic assessments of nuclear energy: on the productionist side, costs of nuclear energy are stable and investments are profitable as long as the profitability is not undermined by political action. The alternative view perceives costs of nuclear energy as unlimited.

(2.) Needs and Production: This dimension stresses different values and needs, on the one hand, and different principles of production for the satisfaction of these values and needs, on the other. The creation of wealth for productionists is a moral imperative. The defining characteristic of an alternative development is the contrasting world view or ideology that questions the emphasis on wealth and material well-being that is prevalent in modern society. Economic growth is a key concept. Considering these core values, proponents of productionist and alternative development have different perceptions of nature and knowledge. Productionists see nature as an ample reserve which should mainly serve the needs of humankind and can be exploited by needs of technological progress. The opposite position stresses the limited resources which demands a careful treatment of natural resources. Human knowledge, particularly in the form of technological progress, is limited and therefore the natural balance should not be disturbed. Consequently, both world-views stand for substantially different concepts of societal productions. Productionists want a continuation of large-scale production. Employment aspects are most important in this concept. From the point of view of proponents of an alternative development a reorganisation of societal production is necessary, which stresses smaller scales and human organisation of work. In this light, the results of production are more important than just high employment rates.

(3.) Utopias of Societal and Political Development: This dimension focuses on different perceptions of social development and grounds for decision-making of social development. Both world-views answer the question about the societal consequence of nuclear energy differently. The notion utopia is not being used here in a sense of an unattainable dream: rather it is a set of beliefs about how society should really work. For productionists, nuclear energy opens up new possibilities, while it is the opposite in the view of alternative politics. The rationality for decisions are also different. Productionists prefer an instrumental rationality and demand a separation of facts and values, and thoughts and feelings. Furthermore, they are prepared to take certain risks. Proponents of an alternative development, in contrast, wish to integrate facts and values, and rationality and emotions. They are much less prepared to take risks which they also see much higher and severe than productionists.

The elements of these opposing world views have been searched for in a content analysis of membership newspapers of employer associations and trade unions. All the statements of the described dimensions - economic concepts, social needs and production, and utopias of societal and political development - are included in an index that indicates the two described world views. Technically, the index can take values between -100 (absolute rejection of alternative aspects and absolute stress of productionist aspects) and +100 (absolute favouring of alternative world views and absolute rejection of productionist aspects). However, since the index is limited by its potential maximum value, the number of cases and the fact that it refers to relatively abstract or extreme concepts, it can be expected that it scores most frequently around 0. Since the index builds upon the fact that statements are mentioned which refer to the different world views, a low value can be caused either by a high number of both alternative and productionist statements, or by not referring to those arguments at all. With all these restrictions in mind the index may give us a rather reliable picture of the attitudes towards a societal development oriented around the technological progress of employer associations and trade unions. However, before I come back to this point I would like to introduce the analysed empirical material and give a more quantitative overview of the attention of the nuclear energy issue.

Empirical material

The empirical study relies on a quantitative content analysis of the membership newspapers of individual trade unions and the peak organisation of the employer associations in Sweden and Germany between the years 1973 and 1990.

Organisational newspapers have several important functions. They promote intra-organisational communication and disseminate the world views of the organisation to their members, the public and other relevant political actors. In short: organisational newspapers are one of the major means in the symbolic struggle of universalising preferred societal world views. Furthermore, I have concentrated also on the membership newspapers, because they offer two methodological advantages: firstly, it is relatively easy to gain access to them, and secondly, they allow a systematic analysis over time.

The debate on nuclear energy in employer associations and trade unions in Sweden and Germany

For the empirical analysis, we can distinguish between different periods of the politicisation of nuclear energy. First, there is the beginning of the politicisation of nuclear energy, when, during the early and mid-1970s, debates on nuclear energy left the expert and administrative sphere and became a public issue. The second period is characterised by a polarisation of the debate between 1976 and 1978, while 1979 and 1980 form the third and fourth period. This period is marked by the accident at Three Mile Island (TMI) in 1979 which demonstrated the danger and possibility of a nuclear catastrophe. In 1980 political actors stabilised or revised their stands. For example, in Sweden a referendum on nuclear energy was conducted in this year. Between 1981 and 1985, the fifth period, nuclear energy was de-politicised. The accident of the nuclear power plant at Chernobyl brought the first direct experience of a nuclear catastrophe to Europe. Both countries have been directly effected by the accident in the Ukraine; 1986 constitutes the sixth period, while 1987 and 1988 is the period of normalisation after the experience of Chernobyl. In this seventh period some re-orientation took place. The last part of this investigation considers the period between 1989 and 1990. The changes in Eastern Europe and the increasing debate of the greenhouse-effect may have led to a re-assessment of nuclear energy and technological progress and economic growth.

Figure 1 and 2 show the frequencies of articles dealing with nuclear energy in the membership paper of Swedish and German employer associations and trade unions.

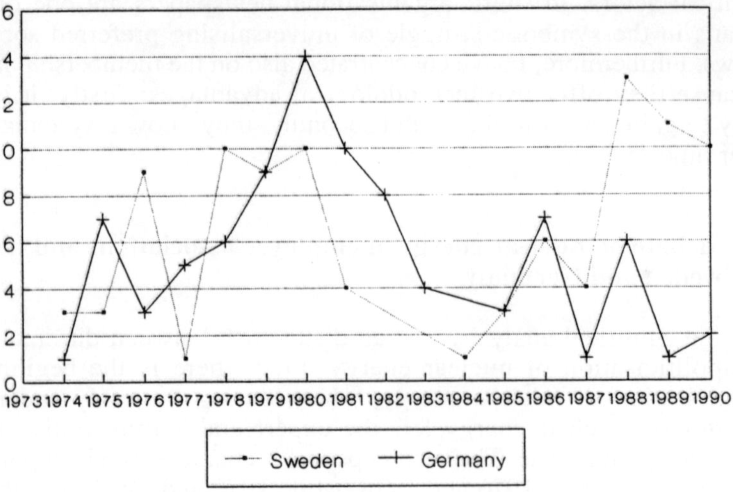

Figure 17.1 The nuclear energy debate of employer associations in Sweden and Germany, 1973-1990

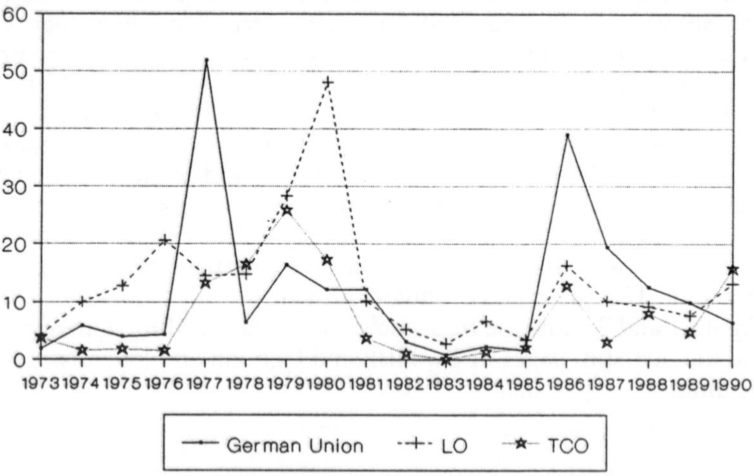

Figure 17.2 The debate on nuclear energy of German and Swedish trade unions, 1973-1990

Figure 1 compares the number of articles of the employer associations in Sweden and Germany. Here we see that the intra-organisational debate in both countries started in the mid-1970s. In Sweden the debate had its first peak during the election in 1976 when the nuclear energy issue was hotly debated and the debate increased again in 1978. However, the accident of the reactor at TMI led only to a slight decrease of articles. During the referendum campaigns the Swedish employer association increased its publication efforts. In the following years the debate faded, but was re-vitalised in 1986, when the reactor accident in Chernobyl happened. The German employer association experienced a similar publication pattern until the late 1980s. However, in the period following Chernobyl the two employer associations differ. While the nuclear energy debate in the German employer papers decreased, the Swedish employer increasingly published articles on nuclear energy. 1987 was the peak year for publishing articles and in the years to come the level have remained high. This behaviour may indicate that Swedish business organisations intend to influence the start of the phasing out program of the government which is due to begin in the mid-1990s.

The debate on nuclear energy in trade unions (figure 2) followed a similar pattern as that of the employer associations. However, the trade unions reacted more strongly to political events and nuclear accidents. For example, the German unions had very high numbers of articles on nuclear energy in 1977. It was during this year that nuclear energy reached the unions for the first time. Articles were controversial and illuminated different aspects, positive and negative, of nuclear energy. However, intra-organisational problems initiated by different attitudes on nuclear energy may have led to a reduction of articles. The TMI-accident led to an increase in the nuclear energy debate again which was to last until around 1981. In the years between 1981 and 1986, there was not much attention given to nuclear energy by any of the trade unions. However, the accident at Chernobyl had a very different impact on the union press which may reflect the general attention given to this accident in both countries (Jahn, 1988). While the German unions substantially increased their publications, the Swedish unions only responded moderately in comparison to the 1970s. Again, similar to the employer associations, the union debate on nuclear energy increased in 1990 in Sweden, while it faded in Germany.

As figure 2 shows we distinguished between the blue-collar unions (LO) and the white collar unions (TCO) in Sweden. The former is closely connected to the Social Democratic Party, while the latter tries to remain neutral on partisan political issues (Wheeler, 1975; Micheletti, 1985). We can also identify this structure in the nuclear energy debate of LO and TCO unions (Jahn, 1992a). While LO unions substantially published at those times when the nuclear energy issue was party politically hot (election 1976, referendum 1980), TCO unions gave the issue the most attention when it was a "news-issue" (accidents at TMI 1979 and Chernobyl 1986). The increase of articles in both unions in 1990 shows that organised interests are arming themselves for a new battle on the phasing-out of Swedish nuclear energy.

These reflections on the organisational debate on nuclear energy in trade unions and employer associations gives only the framework for our analysis of the challenge of technological progress and the undermining of the societal consensus between labour and capital. These aspects will be focused upon in the following section.

The impact of the challenge of technological progress on employer associations and trade unions in Sweden and Germany

Not surprisingly employer associations express a more productionist world view than trade unions. Figures 3 and 4 demonstrate that the attitudes remained quite stable over the analysed time period.

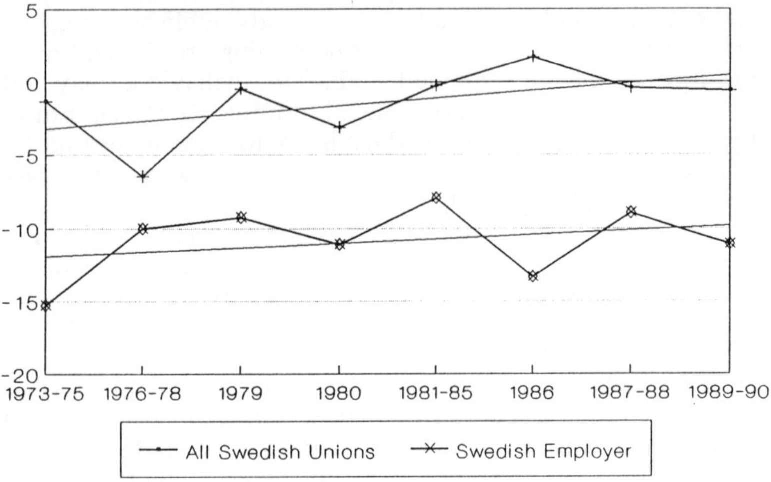

Figure 17.3 Productionist and alternative world views of Swedish trade unions and employer associations, 1973-1990

In particular the employer associations did not change their attitude much. In the first period they started the debate on nuclear energy with a substantial number of productionist statements which support the societal development oriented towards technological progress and economic growth. However, it is also obvious that employer associations stress the productionist world views when the nuclear energy issue is politicised. This is mostly the case in the year 1986 when the accident in Chernobyl occurred, but it is also the case at the time of the referendum. At the time of the referendum campaign, both employer associations and trade unions increased their dissemination on productionist world views. However, the accident at Chernobyl was interpreted differently by the unions and employers. While the former moved away from productionist views, the latter was extremely eager to stress these aspects.

In the last analysed period from 1989 until 1990 the employer association increased the productionist statements again. This may be so, above all, because of the re-vitalisation of the phasing-out debate in Sweden. Further reasons, however, may be the greenhouse effect and the developments in Eastern Europe. The latter in particular may have the effect that material aspects gain in importance again.

The Swedish unions disseminated productionist world views as well, although they did it on a lower level and there was a slight trend towards less productionist world views. In contrast to the employer association, trade unions refer less to productionist world views in periods of accidents of nuclear power plants (1979, 1986), whereas, in periods of political debate (1976-78 and 1980) they increase the stress of productionist world views. In recent years we can also identify a trend towards an increase of productionist world views.

All in all, we may conclude from the presented data that the gap between capital and labour did not increase in Sweden over time, although trade unions do not support a productionist development of society as much as the employer associations; they do not differ qualitatively. Thus we can conclude that the societal consensus between capital and labour has not been fundamentally challenged in Sweden. Is this so in Germany?

Figure 4 presents the data for the debate in employer associations and trade unions from 1973 until 1990.

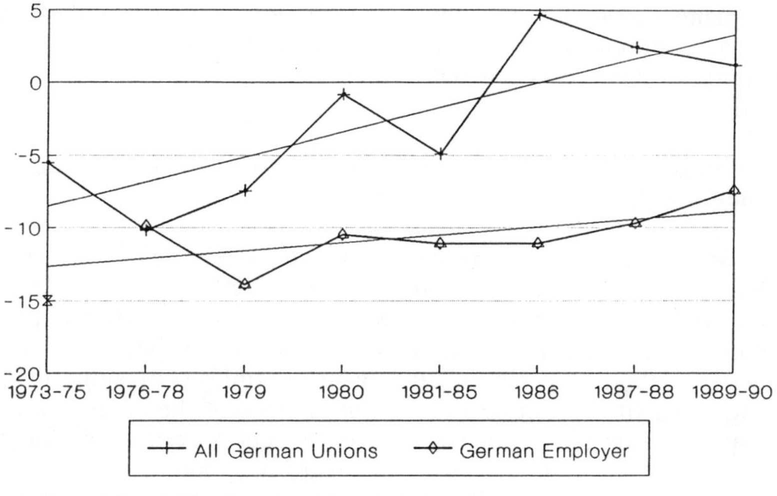

Figure 17.4 Productionist and alternative world views of German trade unions and employer associations, 1973-1990

As in Sweden, the German employer associations kept a rather stable position in their relationship to the productionist world views. The exception is the period 1989-90. In this period they used less productionist statements than in the periods before, although it is important to note that the world view of German employer associations is still strongly guided by productionist aspects. This score is influenced by the surprising abandonment of the reprocessing plant by German enterprises.

Looking at the German trade unions we can identify a trend moving from a productionist idea towards one of alternative world views. Of specific interest was the debate resulting from the reactor accident at Chernobyl when

German unions switched to more critical standpoints, that is towards technological progress and economic growth. However, during the following years the index lost its profile. But it is important to note that the index remained above zero.

Concerning the question about the undermining of the societal consensus on technological progress in Germany, we may conclude that there are some indications that German unions and employer associations differ quite fundamentally on this dimension especially after 1986. However, to conclude that the societal consensus between capital and labour has vanished would be too much; nevertheless, there is a kind of erosion. It is important to note that the debate is not at all at an end. In particular the German unions may turn to a stronger emphasis of productionist world views following German unification.

But it is important to know that the attitudes of the trade unions are not unified. For instance, there are differences between white and blue collar unions, as in Sweden, or between different ideologies of trade unions, as in Germany. In what follows, I wish to analyse whether or not these factors play any substantial role in the different attitudes towards technological progress in trade unions in Sweden and Germany.

The impact of the challenge of technological progress within the trade union movement

As we could see from the section above, in Sweden, in sharp contrast to Germany, there was no increase in conflict between trade unions and employers in terms of their perception of technological progress and economic growth. However, this does not mean that there is no conflict between labour and capital. It is also possible to presume that there is a sharpening of conflict between some factions of the labour movement with capital. If this is true one may even expect a conflict within the unions movement concerning the attitudes towards technological progress.

Since trade unions in Sweden are organised according to blue and white collar workers, it may be feasible to argue that there is a rift between these two unions. However, the classification between blue and white collar unions may not be the major factor for the different perception of technological progress by Swedish trade unions. More important than the membership, the organised sector or industry, or the union structure is the ideology of a trade union (Logan and Nelkin, 1980; Touraine et al., 1987; Jahn, 1993b). Concerning Swedish and German unions it is possible to distinguish five different trade union ideologies (Jahn, 1993b: 154-160).

In Sweden there is an ideological difference between the blue collar LO unions and the white collar TCO unions. While the former unions are strongly committed to a pragmatic social democratic ideology the latter postulate to be non-partisan political trade unions. For German unions we may distinguish between two different ideologies of trade unions:

accommodationist and activist unions. Several empirical studies of German trade unions have been conducted along this line (Bergmann et al. 1976; Markovits, 1986; Weischer, 1988):

There have always been two main groups of unions within the DGB, which we have decided to call activists and accommodationists, roughly corresponding to the more general notions of "radicals" and "moderates". This cleavage, based both on political attitudes and on tradition, has been the principle axis along which intra-DGB politics revolved throughout the post-World War II period. (Markovits, 1986: 20)

This leads us to the analysis of four different types of trade unions:

1 The pragmatic social democratic LO unions which are, as the quotation in the first section illustrated, strongly committed to an efficient national economy and are open to rationalisation.

2 The party-political neutral TCO unions that aim not to get involved in political debates such as nuclear energy (Micheletti, 1985).

The German accommodationist unions which aim to improve the competition of German economy by the promotion of technical innovation and co-operation with capital. In our sample these unions are represented by the Mineworkers' Union (IG Bergbau und Energie) and the Chemical, Paper and Ceramic Workers' Union (IG Chemie, Papier, Keramik).

Finally, we have to do with the activist unions which intend to express an alternative political concept to the existing conditions and which oppose capital most strongly in collective bargaining and in the general political debate. In this sample the activist unions are represented by the Metall Workers' Union (IG Metall), the Printer Workers' Union (IG Druck und Papier), and the Commerce, Banking and Insurance Workers' Union (Gewerkschaft Handel, Banken und Versicherungen).

For the following analysis I summarised the index for the above mentioned trade unions. Figure 5 shows the attitudes of these four types of trade unions.

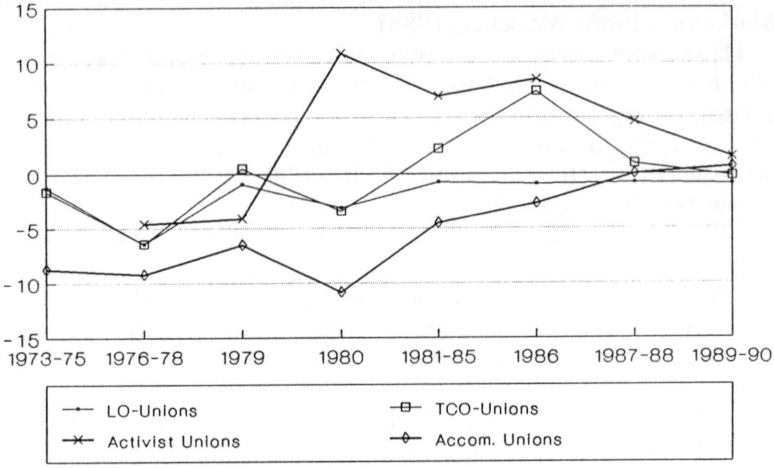

Figure 17.5 Productionist and alternative world views and trade union ideology in Sweden and Germany, 1973-1990

By and large, the development of the attitudes towards productionism shows a trend towards less emphasis on productionist aspects. In the 1970s all analyzed trade unions were strongly committed to industrial values. Particularly in the period of the first polarisation between 1976 and 1978 trade unions were clear about their standpoints. Accommodationist unions were more productionist than activist unions leaving the Swedish trade unions in the middle.

The accident at TMI resulted in bigger changes among Swedish unions than among German ones. LO unions and even more TCO unions changes their attitudes considerably. Productionist statements were less frequently voiced in their membership newspapers. In sharp contrast, German unions did not change their productionist attitudes.

The year 1980 is most interesting for the further development of unions' standpoints. The consequences of the accident at TMI resulted in quite different changes of attitudes among the four ideological types of trade unions. In Sweden we have to recall that a referendum on nuclear energy was conducted in 1980. The referendum campaign involved all political actors and also trade unions took a stance. Even if LO unions campaigned much more clearly for nuclear energy than TCO unions (Jahn, 1992a) as can also be seen of the much higher number of published articles (compare figure 2), the actual stand on the scale of productionist politics is very close between both unions. It becomes also clear, that Swedish trade unions went back to their productionist stand after the shift in 1979. A similar trend towards a stronger emphasis of productionism can also been seen among accommodationist trade unions in Germany. The score was never as productionist as in 1980 and comes very close to the one of employer associations. Probably that is so because the anti-nuclear protest became institutionalised in established politics in Germany. This resulted in established political actors having to take a more decisive stance than before. However, most obvious is the impressive change of attitudes of activist trade unions. In 1980, activist trade unions changed camps. They clearly moved from productionist argumentation towards the alternative side. As we can see from figure 5 this should not change until the present day, although the position is becoming less clear.

The change of the activist unions established a new cleavage between the German unions. It becomes clear that accommodationist unions are still committed to industrial values and productionism while activist unions prefer elements of an alternative world-view. This change makes activist unions a potential alliance partner for new social movements. However, accommodationist unions are openly opposing new social movements as enemies of the social-welfare state.

The years between 1981 and 1985 are characterised by a very low degree of polarisation. As can be seen in figure 2 the debate on nuclear energy in trade union membership newspapers almost disappeared. Therefore, the index scores should not be overestimated in this period.

1986 was characterised by the reactor catastrophe at Chernobyl. The debate on nuclear energy was suddenly re-vitalised. However, apart from the exception of TCO unions, attitudes towards productionism remained astonishingly stable. LO unions hardly change anything. The accommodationist unions did not change position anymore than after the TMI accident. The change of TCO unions in 1986 is surprising. However, this change may be a result of the direct reaction to the catastrophe at Chernobyl. Soon after, TCO unions resumed their more productionist position.

Productionist aspects gained further importance again in the years after 1986. Most interesting, however, is the fact that the gap between accommodationist and activist unions in Germany becomes smaller over the time. This is a result of alternations in both camps. Accommodationist unions moved closer to the activist unions and activist unions gave up their extreme position. This trend continued in face of the German unification. However, it is important to note that the weak score results also from the fact that the nuclear energy debate lost importance within trade unions. Other issues received a much higher priority in Germany.

We can summarise the results by concluding that Swedish trade unions decreased the emphasis placed on productionist values, but without turning to alternative ones. Therefore, the conclusion is that there is a difference between blue and white collar unions in Sweden but this difference comes only to the foreground when the issue is. In normal times both unions support productionist world views. All in all, the difference is not strong enough to be observed as being a fundamental difference between blue and white collar unions. For German unions we may conclude that the challenge of societal development oriented towards technological progress and economic growth formed a cleavage among German trade unions that goes along the lines of accommodationist and activist unions.

Conclusion

The theoretical consideration of the role and status of the belief in technological progress in industrial culture is crucial for the development of society. In particular the political function of being the intermediator for bourgeois' and labour's ideas of the development of society gives special attention to the challenge of technological progress. The whole concept of the institutionalisation of the class conflict or the unity of established political actors such as the labour movement may be questioned. This is the analytical starting point for the comparative analysis in this article.

The empirical analysis of business associations and trade unions in Sweden and Germany demonstrates that the extent to which technological progress is being challenged is rather different in both countries. In Sweden, in contrast to the assumption expressed in the 1970s, technological progress has not been challenged to a larger degree. As a consequence there was no increasing tension between labour and capital or within the trade union movement.

When looking at Germany, we see that on the dimension of productionist/alternative world views the gap between capital and labour has widened. Since the compromise on societal development oriented towards technological progress and economic growth is a basic principle for industrial society, this gap may lead to the conclusion that there are some hints for an undermining of the societal consensus. However, this cleavage does not run very nicely between labour and capital, but instead it shows that there is a cleavage between parts of labour, on the one hand, and capital and other parts of labour, on the other. This conclusion even leads to the fact that we may expect a rising cleavage within the labour movement. This cleavage runs along the lines of "radical" versus "moderate" factions in the labour movement.

The early 1990s have become challenging times for the ideas that criticise productionism as economic security and nationality has reasserted itself throughout Europe. The revolutionary processes in Eastern Europe and the unification of the Federal Republic of Germany may have led to a decrease in the importance of the cleavage analysed in this article. Particularly the standpoints of the proponents of technological progress have been supported by this development and material aspects gained in importance. At the moment, it is difficult to predict whether or not this trend will continue or if the opposition against the dominant industrial culture will become alive again. However, the form and content of such protests can change significantly in that it moves from a protest that emphasises the irrationality of social development and fragmentation of culture by stressing egalitarian values, towards a protest that hopes to interrupt the process of alienation by marginalization of underprivileged groups and stressing national values.

References

Alemann, U. von (1983), 'Krisen der Arbeitsgesellschaft - Katharsis der Interessenvermittlung? Parteiensysteme und soziale Bewegungen im ökonomisch-ökologischen Umbruch', pp. 535-53 in Matthes, J. (ed.), *Krise der Arbeitsgesellschaft? Verhandlungen des 21. Deutschen Soziologentages in Bamberg 1982*, Frankfurt am Main, Campus.

Alexander, J. (1990), 'Between progress an apocalypse: social theory and the dream of reason in the twentieth century', pp. 15-38 in, Alexander, J. and Sztompka, P. (eds.) *Rethinking Progress: Movements, Forces and Ideas at the End of the Twentieth Century*. London, Unwin Hyman.

Alexander, J. and Sztompka, P. (eds.) (1990) *Rethinking Progress: Movements, Forces and Ideas at the End of the Twentieth Century*. London, Unwin Hyman.

Axelrod, R. (ed.) (1976) *Structure of Decision: The Cognitive Maps of Political Elites*. Princeton, New Jersey, Princeton University Press.

Berger, P. and Luckmann, T. (1966), *The Social Construction of Reality. A Treatise in the Sociology of Knowledge*. Garden City, Doubleday.

Bergmann, J., Jacobi, O. and Müller-Jentsch, W. (1976), *Gewerkschaften in der Bundesrepublik. Gewerkschaftliche Lohnpolitik zwischen Mitgliederinteressen und ökonomischen Sachzwängen.* (vol. 1) Frankfurt am Main, Aspekte Verlag.

Blumer, H. (1957), 'Collective Behavior', pp. 127-158 in, Gittler, J.B. (ed.) *Review of Sociology: Analysis of a Decade*. New York, John Wiley.

Brand, K.-W. (ed.) (1985), *Neue soziale Bewegungen in Westeuropa und den USA. Ein internationaler Vergleich.* Frankfurt am Main, Campus.

Cotgrove, S. (1982), *Catastrophe or Cornucopia: The Environment, Politics and the Future*. Chichester, John Wiley & Sons.

Cotgrove, S. and Duff A. (1980), 'Environmentalism, Middle-Class Radicalism and Politics', *Sociological Review*, 28 (2), 333-351.

Dahrendorf, R. (1959) *Class and Class Conflict in Industrial Society*. London, Routledge & Kegan Paul.

Dalton, R. and Kuechler, M. (eds.) *Challenging the Political Order*. Oxford, Polity Press.

Eder, K. (1990a) 'Counterculture movements against modernity. Nature as a new field of class struggle', *Culture and Society*, 7, 21-47.

Eder, K. (1990b), 'The cultural code of modernity and the problem of nature: a critique of the naturalistic notion of progress', pp. 67-87 in, Alexander, J. and Sztompka, P. (eds.) *Rethinking Progress: Movements, Forces and Ideas at the End of the Twentieth Century*. London, Unwin Hyman.

Erd, R. and Scherrer, C. (1985), 'Unions - Caught between Structural Competition and Temporary Solidarity: A Critique of Contemporary

Marxist Analysis of Trade Unions in Germany', *British Journal of Industrial Relations*, 23 (1), 115-131.

Esser, J. (1982), *Gewerkschaften in der Krise. Die Anpassung der deutschen Gewerkschaften an neue Weltmarktbedingungen*. Frankfurt am Main, Suhrkamp.

Eermann, R. and Jamison, A. (1991), *Social Movements: A Cognitive Approach*. Oxford, Polity Press.

Flam, H. (in colabouration with Andrew Jamison) (1992), 'The Swedish Confrontation over Nuclear Energy - A Case of a Timid Anti-Nuclear Movement', in, Helena Flam (Hrsg.), *States and Anti-Nuclear Movements*, Edinburgh, Edinburgh University Press..

Gamson, W.A. (1988) 'Political Discourse and Collective Action', pp. 219-244 in, Klandermans, B., Kriesi, H. and Tarrow, S. (eds.) *From Structure to Action: Comparing Movement Participation Across Cultures*. Greenwich (Conn.), JAI.

Gottweis, H. (1986) *Zur Politisierung des Energiesektors. Ein internationaler Vergleich. Vstereichische Zeitschrift für Politikwissenschaft*, 15 (1), 43-59.

IGBE, (1992) *Energiepolitik: Positionen, Beschlüsse, Fakten*. Bochum.

Israel, J. (1978), 'Swedish Socialism and Big Business'. *Acta Sociologica*, 21 (4), 341-53.

Jahn, D. (1988), 'Tschernobyl und die schwedische Energiepolitik'. *Österreichische Zeitschrift für Politikwissenschaft*. 17 (1), 43-51.

Jahn, D. (1992a), 'Nuclear Energy in Swedish Trade Union Press from 1973-1989'. *Nordicom Review of Nordic Mass Communication Research*, (1), 23-33.

Jahn, D. (1992b), *Nuclear Power, Energy Policy and New Politics in Sweden and Germany. Environmental Politics*, 1 (3), 383-417.

Jahn, D. (1992c), 'The Role of Organizations in the Establishment of Ecological Consensus in Industrialized Countries', in, Svedin, Uno und Aniansson, Britt (eds.) *Society and the Environment,* Dordrecht, 213-228.

Jahn, D. (1992d), *Utopia and Ideologies of Environmental Movement Organizations in Sweden and Germany*, paper delivered at the First European Conference on Social Movements, Berlin, October 29-31, 1992.

Jahn, D. (1992e), *The Impact of New Politics in Sweden and Germany*, paper delivered at the 88th Annual Meeting of the American Political Science Association, Chicago, September 3-6. 1992.

Jahn, D. (ed.) (1993) 'Europa im Umbruch: Neue soziale Bewegungen im Schatten des europäischen Binnenmarktes.' Special issue of the Forschungsjournal *Neue Soziale Bewegungen*, vol. 6, no. 1.

Jahn, D. (1993a), 'The Rise and Decline of New Politics and the Greens in Sweden and Germany: Resource Dependence and New Social Cleavages', *European Journal of Political Research*, (in print).

Jahn, D. (1993b), New Politics in Trade Unions. Aldershot, Dartmouth.

Jamison, A., Eyerman, R., Cramer, J. and Lfssve, J. (1990), *The Making of the New Environmental Consciousness. A Comparative Study of the*

Environmental Movements in Sweden, Denmark and the Netherlands, Edinburgh, Edinburgh University Press.

Kitschelt, H. (1983), *Politik und Energie. Energietechnologiepolitiken in den USA, der Bundesrepublik Deutschland, Frankreich und Schweden.* Frankfurt am Main, Campus Verlag.

Kitschelt, H. (1984), *Der ökologische Diskurs. Eine Analyse von Gesellschaftskonzeptionen in der Energiedebatte.* Frankfurt am Main, Campus Verlag.

Kitschelt, H. (1986) 'Political Opportunity Structures and Political Protest: Anti-Nuclear Movements in Four Democracies'. *British Journal of Political Science*, 16 (1), 57-85.

Klandermans, B. (ed.) (1989), *International Social Movement Research.* 2 vols., Greenwich (Connecticut), Jai Press.

Lindberg, L. (1977), 'Energy Policy and the Politics of Economic Development'. *Comparative Political Studies*, 10, 355-82.

Logan, R. and Nelkin, D. (1980), *Labor and Nuclear Power. Environment,* 22 (2), 6-13 and 34.

Markovits, A. S. (1986) *The Politics of the West German Trade Unions. Strategies of Class and Interest Representation in Growth and Crisis.* Cambridge, Cambridge University Press.

Martin, A. (1984), 'Trade Unions in Sweden: Strategic Responses to Change and Crisis', pp. 191-359 in, Gourevitch, P. et al. (eds.) *Unions and Economic Crisis*. Boston, Allen and Unwin.

Micheletti, M. (1985), *Organizing Interest and Organized Protest. Difficulties of Member Representation for the Swedish Central Organization of Salaried Employees* (TCO), Stockholm, Department of Political Science.

Mlller-Rommel, F. (ed.) (1989), *New Politics in Western Europe. The Rise and Success of Green Parties and Alternative Lists*. Boulder (Colorado), Westview Press.

Offe, C. (1981), 'The Attribution of Public Status to Interest Groups', pp. 123-58 in, Berger, S.D. (ed.) *Organizing Interests in West Europe.* Cambridge, Cambridge University Press.

Offe, C. (1985), 'New Social Movements: Challenging the Boundaries of Institutional Politics', *Social Research*, 52 (4), 817-868.

Perlman, S. (1966), *A Theory of the Labor Movement*. (original edition 1928) New York, Kelley.

Perrow, C. (1986), *Complex Organizations A Critical Essay*. (3rd edition) New York, Random House.

Rammert, W. (1992), 'Wer oder was steuert den technischen Fortschritt? Technischer Wandel zwischen Steuerung und Evolution'. *Soziale Welt*, 43 (1), 7-25.

Rucht, D. (ed.) (1991), *Research on Social Movements. The State of the Art in Western Europe and the USA*. Frankurt am Main, Campus and Boulder, Westview.

Sabel, C.F. (1981), 'The Internal Politics of Trade Unions', pp. 209-44 in,

Berger, S.D. (ed.) *Organizing Interests in West Europe*. Cambridge, Cambridge University Press.

Sahr, R.C. (1985), *The Politics of Energy Policy Change in Sweden*. Ann Arbor, University of Michigan Press.

Sarcinelli, U. (1987) *Symbolische Politik. Zur Bedeutung symbolischen Handelns in der Wahlkampfkommunikation der Bundesrepublik Deutschland*. Opladen, Westdeutscher Verlag.

Schmitter, P.C. (1988), 'Five Reflections on the Welfare State', *Politics and Society*, 16 (4), 503-515.

Snow, D., Benford, R.D. (1988), 'Ideology, Frame Resonance, and Participant Mobilization', pp. 137-196 in, Klandermans, B., Kriesi, H. and Tarrow, S. (eds.) *From Structure to Action: Comparing Social Movement Research Across Cultures*. Greenwich, Jai Press.

Snow, D., Rochford, E.B., Worden, S.K., Benford, R.D. (1986), 'Frame Alignment Processes, Micromobilization, and Movement Participation', *American Sociological Review*, 51 (4), 464-481.

Tilton, T. (1990), *The Political Theory of Swedish Social Democracy. Through the Welfare State to Socialism*. Oxford, Clarendon Press.

Touraine, A. (1971), *The Post-Industrial Society. Tomorrow`s Social History: Classes, Conflicts and Culture in the Programmed Society*. New York, Random House.

Touraine, A., Wieviorka, M. and Dubet, F. (1987), *The Workers' Movement*. Cambridge, Cambridge University Press.

Weick, K.E. (1979), *The Social Psychology of Organizing*. (2nd edition), Reading (Massachusetts), Addison-Wesley.

Weischer, C. (1988), *Kritische Gewerkschaftstheorie. Ansätze einer Neuorientierung*. Münster, Verlag Westfälisches Dampfboot.

Wheeler, C. (1975), *White-Collar Power. Changing Patterns of Interest Behavior in Sweden*. Urbana, University of Illinois Press.

This article discusses some preliminary findings and hypotheses from the research project 'The Institutionalisation of Ecological Issues in Modern Societies' supported by the Swedish Council for Planning and Co-ordination of Research (forskningsredsndmnden; grant: Dnr 890721:2 A 8-5/105) in the framework of the World Commission on Environment and Development (Brundtland Commission) and the Minister of Research and Technology of the Federal Republic of Germany (SWF 0048 0). The aim of this project consists of analysing of the expression and character of ecological politics (new social movements and Green parties) and the reception, integration, and reformation of ecological standpoints in established politics (parties, interest groups of labour and capital, parliaments, mass media). It is conducted at the Department of Political Science at the University of Gothenburg (Sweden) and the Department of Sociology at the University of Paderborn (Germany). I wish to thank Gunilla Augustsson, Gunnel Vessman, Maria Skattner, Doris Wilting, Volker Eickhoff and Sabine Neudorf for the data collection and processing.

New politics is characterised by simultaneous stressing of (new) social issues such as environmental pollution, peace, participation right, etc., (new) modes of collective action like demonstrations, boycotts, spontaneity, etc., (new) values of self-realisation, personal autonomy and identity, and new social actors who are young, highly educated, and more removed from productionism, coming together in new social movements, and organising themselves in Green parties, etc. In political science, these aspects have been the focus of great attention in the last decade in the context of party research.

The unions considered in this investigation are the following Swedish blue collar unions: LO (Landsorganisationen i Sverige), Confederation of Swedish Blue Collar Workers; Metall (Svenska Metallindustriarbetarenfvrbundet), the Swedish Metal Workers' Union; SKAF (Svenska Kommunalarbetarefvrbundet), the Swedish Municipal Workers' Union; SF (Statsanstdlldas Fvrbundet), the Swedish Union of State Employees; SEF (Svenska Elektrikerfvrbundet), the Swedish Electricians' Union; GF (Grafiska fackfvrbundet), the Graphic Industry Workers' Union. The Swedish white collar unions are: TCO (Tjdnstemdnnens Centralorganisation), the Swedish Confederation of Professional Employees; SIF (Svenska Industritjdnstemannafvrbundet), the Swedish Union of Clerical and Technical Employees in Industry; SKTF (Sveriges Kommunaltjdnstemannafvrbund), the Swedish Union of Local Government Officers; ST (Statstjdnstemannafvrbundet), the Union of Civil Servants. The German trade unions are: DGB (Deutscher Gewerkschaftsbund) Confederation of the German unions; IGM (IG Metall), the Metal Workers' unions; VTV (Gewerkschaft Öffentliche Dienste, Transport und Verkehr), the Public Service and Transport Workers' union; IGD (IG Druck und Papier, in 1990 IG Medien) the Printer Workers' unions; HBV (Gewerkschaft Handel, Banken und Versicherungen), the Commerce, Banking and Insurance Workers's unions; IGC (IG Chemie, Papier, Keramik), the Chemical, Paper and Ceramic Workers' unions and IGB (IG Bergbau und Energie), the Mineworkers' union. All the unions of this investigation organise approximately two thirds of the respective membership.

18 Three cultures, one model: The construction system dynamic model in three European countries

Peter Fleissner

Summary

The paper describes personal experiences in interdisciplinary scientific work under different cultural contexts. As an example the construction of extended macroeconomic models was chosen. The method applied was Forresters Systems Dynamics. Compared with the econometric approach and the input-output-method it can be learned rather easily and quickly.

All the models were constructed between 1987 and 1992 in interdisciplinary teams for Austria, for East-Germany, and Slovakia. The teams consisted mostly of scientists who were citizens of the country modelled, and most of the work was done in the respective country under investigation. So the author who is of Austrian background and was in charge of all the teams had the opportunity to compare countries with different social order (capitalism, dying socialism, early stages of a reform economy), with languages equal and different to his mother tongue, with different scientific and political cultures, with different degrees of experience in modelling, and with groups of specialists of different fields etc. Additional partial experiences in System Dynamics could be gained by him in the US and the FRG.

The paper deals with a short description of the method of Systems Dynamics applied, the basic features of the models and its extensions. The setups of the studies are compared with respect to its institutional, organizational, financial, and technical base, the formal and informal structure of the teams, misunderstandings, conflicts and harmony, its dynamics and its degree of integration, the relationship of the team to the scientific environment and the degree of acceptance and use of the results of the studies.

The final chapter deals with the pros and cons of smaller models compared with large ones. Two methods are proposed which could make the understanding and evaluating of large models easier.

1 Introduction

About twenty years ago the author of this paper was trained by Dennis Meadows and others in System Dynamics (SD). At these days this method was well known because of the world wide distribution of the books "World Dynamics" by Jay Forrester[1] and the best-seller "The Limits to Growth" by Meadows and Meadows[2] et al which were based on SD. For the first time an environmental breakdown of our world was predicted within the next one hundred years. In 1973 under the impression of Forrester's and Meadow's writings the Austrian Academy of Sciences founded the Institute for Socio-Economic Development Research. The author became the first scientific worker hired. There the SD approach should be applied to Austrian affairs, and a kind of societal early warning system was to be implemented. At first a large SD-model[3] was constructed for the Austrian health care system but it was more than that. The model not only included the health care system itself but tried to link this subsystem with the demographic, economic, social and political strata of Austrian society in a dynamic, quantitative and partly empirical way. Nevertheless the SD-model was just one small part (about 5 per cent of the resources were spent for model construction) of a larger research project where theoretical aspects of the health care system on politics, economics, environment, demography, psychology, history and comparative system research were elaborated in detail. 1975 this project was published in a two volumes book[4], shortly after the synchronous international economic crisis. Therefore the chances to implement the results in Austria were very poor. The performance of the economy was bad, there was a drawback in tax collection, governmental budget ran into deficits which could not be levied up to now.

At this time the authors interest in SD developed. Under the impression of high brow mathematical and statistical methods used before in econometric model building (the most powerful method for predicting the economy) SD was seen as a not enough precise method lacking any empirical base. The SD-approach used in "The Limits to Growth" was criticized heavily by a majority of scientists partly because they could show that small changes in the parameters of the model led to totally different results, partly because of the overwhelming success of this book. But soon a change in the assessment of econometrics evolved. It could be shown that there was no country in the years 1974 and 1975 where econometric models had not predicted figures largely deviant (too optimistic) from the actual growth rates[5]: The former thrust into the predictive power of econometric models dwindled away and changed into a general anti-modelling attitude. Aware of the weaknesses of both methods the author changed his focus of interest towards a more general agenda, and - to avoid competition - he tried to define a (for Austria) new branch of science: He called it "Social Cybernetics", a combination of verbal analysis of subsectors of society and mathematical modelling, and based his academic career on it. After some difficulties with bureaucracy, questions of competence in the academic world, and some delays related to it finally he became the first formally announced university teacher in "Econometric Social Cybernetics" in Austria. In the meantime teams where the author was a member produced a lot of qualitative and verbal analyses[6] of subsectors of society, but still they tried different mathematical methods and applied them to societal problems, starting from Game Theory up to Input-Output-Analysis. They have studied the Austrian educational system, problems of new technologies, diffusion and impact of microelectronics in Austria, of industrial robots[7], of flexible manufacturing systems in small and medium enterprise etc.

2 Three modelling efforts

The Austrian model

It became 1986 when there were rumors in the scientific community about a mathematical model of the US economy which should be better than any other before. It should display Kondratieff waves, long waves with a period of 45 to 55 years. This was the next opportunity the author was made aware of Jay Forrester's work. By chance he got the opportunity to study this model at its place of origin, at MIT, Cambridge, Mass. One goal of his stay was to learn and to develop the basics for a similar model of the Austrian economy. Unfortunately it was impossible for him to proof the rumors. At this time no comprehensive SD-model of the US economy was available in a published version although the starting year of its construction was said to be before 1980. It existed as a computer program only but there were contradicting statements about its behavior. In the authors opinion the model was unstable. Where actual results of the model had actually been reported by the authors they were just results of submodels, not the result of the comprehensive set of equations.

Back home in Austria the author had a difficult time to develop a different approach to Forrester's one which promised to end up in a working model within two years time. Compared with Forrester the Austrian team tried a methodology which differed in three aspects:

- First, they did not start with a set of detailed submodels as Forrester has done (he was guided by the individualistic doctoral programs at MIT, each of the students became responsible for one submodel). They tried to create the basic structure by a cooperative effort. All the team discussed and defined the basic structure of the comprehensive model. Submodels were developed afterwards as a result of the split of the comprehensive model. Details could be added but every member of the team was aware about the overall structure. In short, they applied a top-down approach compared with Forrester's bottom-up method.

- Second, they did not do without empirical estimation of the model parameters. They collected time series as far as available and used rule of thumb methods to determine values of "hidden" parameters they were not able to estimate precisely by econometric methods because of the lack of appropriate data (e.g. the speed of adaptation of entrepreneurs for a change in capacity utilization under the conditions of a chronic lack of demand).

- Third, after having finished a submodel with respect to the implementation of all the feedback loops considered and after having estimated all the parameters needed, they started a test run to check the precision of ex-post fit of the simulation model. This test is not unnecessary as somebody could argue eventually. It is not trivial that the simulation fits into the actual trajectories of empirical data. Sometimes error amplification can be a problem and can produce considerable differences between actual and simulated data. In such cases they applied "fine-tuning" of the submodel by multiplying a constant close to one with some of the sensitive equations. The constant was varied such as to minimize the absolute sum of the plain or squared differences between simulated and actual values. Not until they had applied this procedure they started to integrate the submodels. But during this process a strange event happened: After they had added the foreign trade sector to the model (only the government sector remained to be included) it became extremely unstable which would have been disastrous for their efforts. They were lucky when the model stabilized again after they had integrated the last submodel, the government sector.

Finally, the model consisted of 430 equations (partly nonlinear, partly very simple ones). 33 out of them were estimated by OLS, the parameters of 7 equations were determined by trial and error and best fit. There were 20 exogenous variables included. The data base was 1964 to 1987, and the simulations were run to the year 2012.

The ex-GDR model

In the summer of 1989, the year of the publication[8] of the Austrian model in Vienna, many GDR tourists who spent their holidays outside GDR in other socialist countries used the opportunity to emigrate via borders free of the destroying installations of the "iron curtain". This was the first symptom of the imminent end of the German Democratic Republic. After December 1989 (when some political activists in GDR proposed a democratic reform of a nevertheless socialist country) it became evident that the integration of the GDR into the Federal Republic of Germany will be the way into the future. In May 1990 in East Berlin the author had the opportunity to discuss the effects of a change of the social and economic system with colleagues of the

Academy of Sciences of the GDR. He proposed to try a SD-model to assess the different scenarios in a structured way. Within one afternoon they created a simple model structure which reflected the main features of the model they developed later on. In summer 1990 they founded a working group to construct a comprehensive model. It was finished in summer 1991 and it was published as a book[9] in February 1992. The model consisted of 206 statements. The simulation period reached from 1989 to 1996. Only one equation was estimated econometrically.

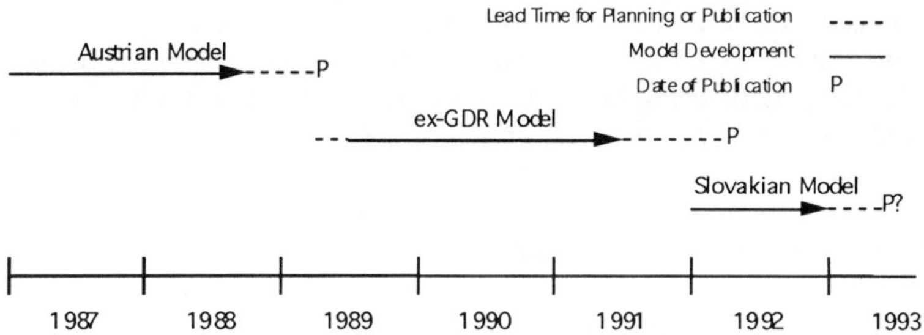

Figure 18.1 Time schedule of model construction

The model of the Slovakian Republic

In Summer 1991 the author applied for a research grant at the Austrian Ministry of Science and Research to study the transformation of our neighbor country Czecho-Slovakia towards a so-called market economy. Czechoslovakia was at that time already written with a dash between the two names of the republics. Separatist ideas grew at this time. When the author looked for a team in Bratislava, the capital of the Slovakian Republic, all the curricula of the potential members of the team he received showed the common mother tongue "Slovakian". With one Hungarian exception each of the members stressed their Slovakian origin. They got a grant by the ministry and started with the modelling process in January 1992. In the meantime after half a year they have finished an interim report[10]. The project should be finished at the end of this year, or maybe in Spring 1993.

System dynamics in a nutshell

In this paragraph a short description of System Dynamics is presented. There are three main ideas behind this method:

- First, all the events of real life must be measurable in a more or less precise way (quantification principle).
- Second, there must be variables which represent stocks, and others

which represent flows (bathtub principle).

- Third, for a more interesting dynamics of the model there must be some feedback loops (feedback principle).

Because of the considerable progress in electronic data processing the user friendliness and ease of use of System Dynamics software have been improved by orders of magnitude.[11] By the development of STELLA, a System Dynamics package with a powerful graphic interface (see e.g. figure 2), now small models can be constructed in very short time.

The necessary learning efforts have been reduced heavily. The software assists the user by offering her/him four essential structural elements (beside other modelling tools) which can be drawn as a single object directly on the screen:

- Stocks (visualized by a rectangle)
- Flows (visualized by a valve or a water tap)
- Auxiliary Variables (visualized by a circle) and
- Causal Influence (visualized by an arrow)

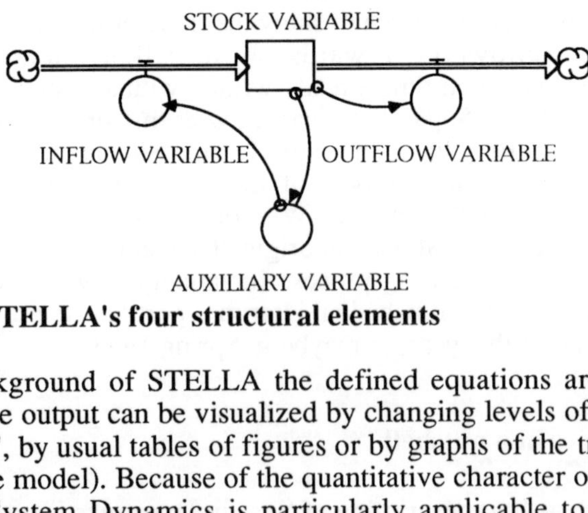

Figure 18.2 STELLA's four structural elements

In the background of STELLA the defined equations are ordered and structured. The output can be visualized by changing levels of the "water" in the "bathtubs", by usual tables of figures or by graphs of the trajectories (the solution of the model). Because of the quantitative character of certain social phenomena System Dynamics is particularly applicable to demographic, economic and/or social processes.

Figure 3 shows one example of a typical model of a production system where capital, labour, potential and actual output are linked together. By modelling of the old fixed capital stock and the new fixed capital invested separately, and by keeping its related technical parameters (average and marginal ones) separate as well, technical change can be reflected in a convenient way.

Another generic element common to each of the models are feedback loops. Figure 4 shows one example of a "positive" feedback loop, the inner mechanics of the break down of the economies of many of the former socialist countries. Positive feedback loops amplify changes in the starting conditions. Is there a negative change in the demand for labour, this negative change will be amplified. A cumulative behavior of the variable is the outcome, either in a negative way as in this example, or in a positive way (e.g. the increase of a population at high fertility rates).

Negative feedback loops are used to generate goal seeking behavior of the model. If there are delays in the course of the negative loop, oscillations (increasing or decreasing in amplitude) may arise.

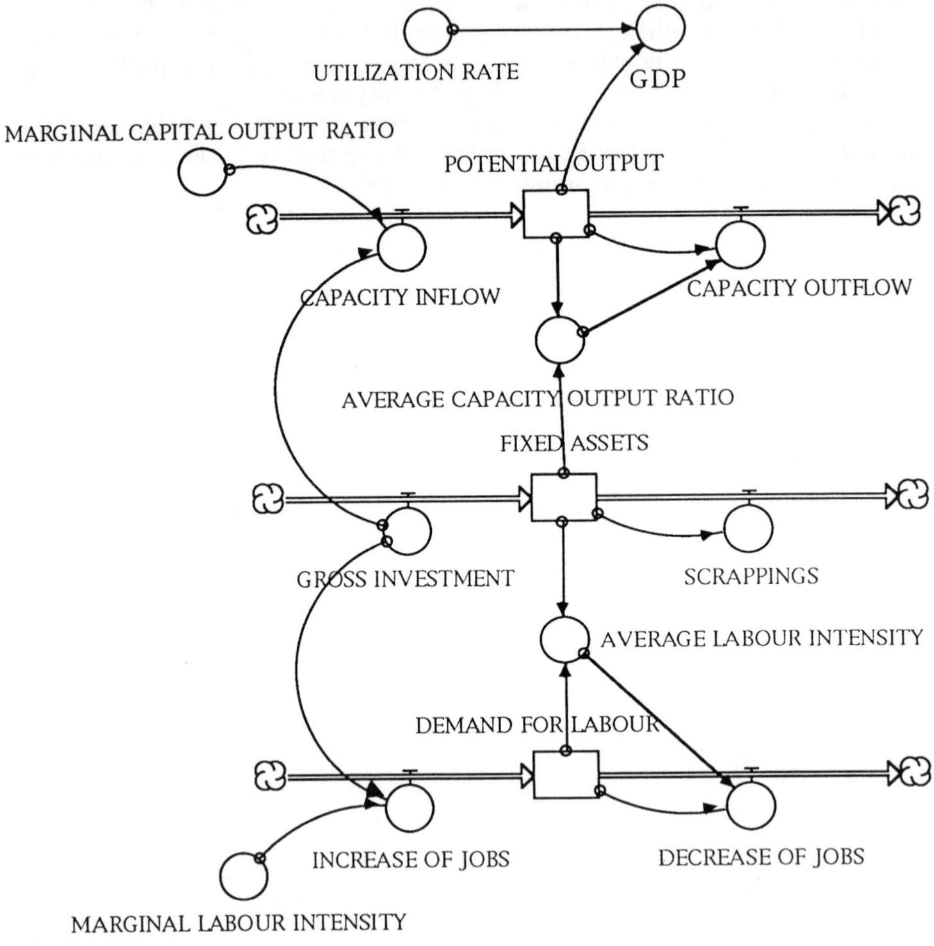

Figure 18.3 Production function

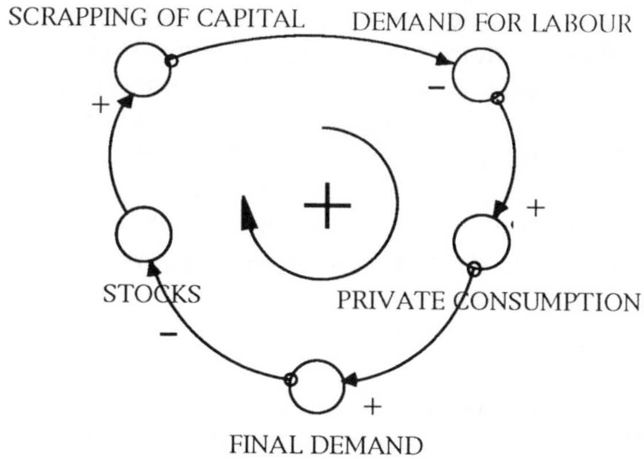

Figure 18.4 Positive feedback loop of the downturn

4 Comparison of the modelling setup and the modelling processes

Financial resources

The Austrian model was financed completely by the Austrian National Funds for Promoting Scientific Research by about 80.000 ECU, the Slovakian study was financed by the Austrian Ministry for Science and Research (Funds for Eastern Europe) by about 100.000 ECU, while the East-German model was not able to attract any financial support by public sources although they tried hard at the Foundation of the Austrian National Bank. In fact the team offered a presentation of the Ex-GDR-model to the German Ministers Möllemann and Waigel but they never received an invitation. So they did their research by the help of exchange programs for scientists. Much of the work remained simply unpaid.

The Range of the Models

While in the case of Austria an economy was described which had a long history in capitalistic development, and many of the statistical data needed could be already found in Statistical Yearbooks and other publications, it was much more difficult to find appropriate data for the situation in East Germany and in Slovakia. Much more expert (or sometimes non-expert) opinions were used as input in the latter cases than for the Austrian model. Therefore it was not possible to use econometric methods for GDR and Slovakia to that large extent as in the Austrian case. The structural changes were most severe in the GDR, they are smaller in Slovakia, and they are minimal in Austria.

All the models consist of the traditional economic sectors like production, final demand, labour market, government and social security, and foreign final demand, labour market, government and social security, and foreign trade. Besides them a demographic sector was included in each of the models. For Austria and East-Germany the discussed problems were limited to economic, social and demographic ones. For Slovakia it is planned to investigate environmental problems as well as they are much more severe and are much more present in the mass media now. In addition they try to develop a monetary and credit submodel which will connect the "real" (physical) side of the economy with its monetary aspects.

The Size and Qualification of the Teams

Looking backward it can be seen that the size of the teams was considerably equal. They consisted of 8 to 9 persons. While in the Austrian team of scientists there was no woman, the contribution of a female secretary to the graphical surface of the computer model was of particular importance for the acceptance of the model by the media (see figure 5 for our use of the "ship-metaphor" to visualize the economic state of a country dynamically).

Figure 18.5 The ship metaphor of the Austrian economy

Three controlling variables can be identified and directly controlled by the mouse at run time, the steering wheel (changes in the tax pressure to Austrian tax payers), the level of the anchor (working hours per week), the angle of the sail (amount of cumulated public investments). Three elements symbolize the performance of the economy: the position of the sun (rate of growth of Gross Domestic product), the position of a cloud (the more to the right the higher the rate of inflation) and the level of the sea (rate of unemployment - not to be seen in figure 5). These three elements are moving around the screen controlled by the full set of equations of the Austrian model. In the GDR-team three women (economists) had participated while in the Slovakian team only one women takes part. She is specialized in collecting data. The qualifications were most homogeneously spread in the GDR-team. Besides the author all of them were economists of different kind. The heterogeneity of the Austrian team was rather high: two physicists, one computer scientist, one psychologist, one electrical engineer, one statistician, one economist and one political scientist. The majority of the Slovakian team are economists, besides one statistician, one political scientist and (partly) one psychologist.

Organization of labour

None of the teams had ex-ante experience with System Dynamics. So it was necessary to give some introduction into the method and the software used. Usually the author used the Macintosh and a black and white display of the computer screen to teach STELLA by means of simple examples of simulation models. The reaction to his presentation was different. Each member of the Austrian team tried to learn STELLA himself and - by neglecting possible effectivity increases by a higher degree of division of labor - produced its own special submodel he was responsible for. They liked to work as all-rounders. They studied economic theory, they collected data, they estimated the necessary parameters, they fine-tuned their models, discussed the output and modified their submodels if necessary. The Slovakian team showed a different behavior. They used their different qualifications to burden the specialized member with specialized work to be done most efficiently by her/him. Now the computer expert is doing all the programming under MS-DOS and PROFESSIONAL DYNAMO PLUS, the mathematician is burdened with creating intelligent formulae for the behavioral equations, the statistician is collecting all the necessary data for the complete team, the official head of the group (they are salaried mostly by the same institute) feels himself responsible for "soft" and "hot" political questions and does the necessary arrangements for meeting places, notes, copying etc. The outcome of such a situation is that political questions are not discussed in detail by all the members of the team but the responsibility seems to be shifted towards the person with the highest rank in the organizational hierarchy. To counteract this tendency the author tried to insist in plenary sessions (one day twice a month) and slowly more general interest and a higher participation rate in discussions can be seen. Similar behavior is shown in the case of conflicts. As a rule contradictory problems are decided by the person of highest rank, not by a coordinated and lengthy process of commonly shared discussions as it is done in the Austrian team. At the same moment the Austrian team showed a high degree of self-organization. The level of this parameter was low in the beginning of the work with the Slovakian and GDR-team but increased considerably over time. It seems as if there was a lack of direct scientific interaction and critics by arguments between social scientists in the former socialist countries.

Similar features can be reported about informal relationships after work. While the Austrian way of life usually is accompanied by a high degree of informal interaction there was considerable interaction with the team in the former GDR, but nearly no interaction with the team of Slovakia. This may be due to language problems because they have decided to use English as our language of communication. This decision should equilibrate the burden a foreign language exerts to the individuals.

Hardware and software

Not only differences in human languages created problems but computer languages and different hardware as well. In the Austrian team they were well equipped with Macintoshes. Each member had access to a Mac. Because of the lack of Macintosh in the former socialist countries the author was urged to shift away from the platform he worked with. Therefore he used to translate PROFESSIONAL DYNAMO PLUS Programs (running under MS-DOS) into STELLA diagrams first. But after the 15th revision he became tired of this job. So he shifted directly to a MS-DOS machine himself but still he found that his virtuosity in creating models was reduced by this change.

The lack of a (fast and silent) laser writer during the first half of the GDR-project was a bottleneck to the efficiency of our work. Although in principle one could read all the necessary information about the model from screen, in practice this is a very cumbersome activity. Computer printouts still can be studied and evaluated much easier than information displayed on screen.

Policy implications

The three models attracted the public awareness in a different way. The Austrian study gained some attention in computer journals, and was discussed at scientific meetings, between economists and interested journalists. The team was not successful to sell the forecasts every quarter of the year to an Austrian newspaper. To illustrate possible applications in the book they dealt with the consequences of a full membership of Austria in the European Common Market, investigated some effects of a reduction of working hours per week, and computed one scenario with an increased speed of technical change. But as far as they know the model was never used in practice to improve economic policy. They tried hard to make the book more attractive to the reader. A computer disk was added to it where every owner of a Macintosh PC could play around with a software developed by the team (see figure 5). Within a period of three years no more than 250 copies of the book were sold only. Maybe they would have been better off by writing software running under MS-DOS because of the higher market share of DOS-machines. But at that time no such user-friendly software like HYPERCARD was available. Now TOOLBOOK could be used under MS-DOS with nearly the same features as HYPERCARD. It includes colored graphs in addition.

The first forecasting results of the East-German model were immediately published in a prominent West German newspaper (two articles in Wirtschaftswoche, some critical remarks could be found in Frankfurter Allgemeine Zeitung). They deviated considerably from other forecasts which were much more optimistic than ours. Unfortunately ours proved to be the more correct ones. About 500 copies of the book were sold during the first months after publishing. Unfortunately no follow-up study could be organized to modify and to improve the model continuously over time up to now.

The Slovakian model is the one which may be the most promising for use in economic policy. There is some hope to be able to present the model to economic decision makers of the Slovakian Republic in fall 1992. The goal is to implement the model as a continuously used tool in the decision making process. Different scenarios of possible economic, social and demographic policy could be tested by the model. If there are real data available on the effects of the actual policy implemented they have to be compared with the models results. If there are differences between simulation and reality the model has to be modified and improved. By this interactive process we hope the model users will be able to get a deeper understanding of the economic and social processes. We do not think that in a period of such drastic changes it is possible at all to create "perfect" or even "correct" models. Human decision makers modify the "laws" of the economy nearly every day. There is no hope for "precise" forecasts: But an improved understanding of what is going on could help to reduce the costs of social change.

5 Medium scale versus large scale models

A comparison between medium scale models like ours (200 - 500 equations) and really large ones (several thousand equations) can be done along different dimensions. Just to name some of them: degree of aggregation, inputs needed, output gained, user-friendliness in updating the model, comprehensibility of the model, simplicity of use in decision making. In the following the author will discuss some of the pros and cons of larger versus smaller models.

Against universal models

The degree of aggregation and - correlated to it - the degree of abstraction of the model has to be fixed according to the number of person-years of experts and technical staff available. This number is a non-decreasing function of money resources. There is a tradeoff between the degree of detail (vertical scope) and the range of the model (horizontal scope). In the authors opinion there is no such thing like a universal model. The models are useful only if they are specialized to a high extent. With limited manpower it is not possible to construct one large macromodel which refers e.g. not only to the overall national demographic, economical, social and political indicators as well as to the respective regional indicators at the same time. Such a model cannot be finished in due time. Therefore one has to decide if one is in favor of a rather small model which is limited to one aspect of reality only (e. g. the political stratum), and to formulate such a model in detail (high vertical scope), or to prefer a highly aggregated and abstract model which is in fact the reunion of rough submodels of different aspects of reality (high horizontal scope). While the first type of models can answer questions of one layer of real world only but these answers can be very detailed, the second one can give informations on the effects of the relationships between different layers of the real world. While the first type of model is a model within the range of one scientific discipline, the second type needs interdisciplinary efforts. An example shall clarify this point: The first type could be an input-output model with several hundred economic activities. Only economists are needed. The second type could be a multi-layer model where population dynamics, migration, the economy, the health care system, and health care politics are modelled and combined into one single structure. One can see that there will be no success without a team which covers all the disciplines needed. In the latter case some time is needed to create a common understanding of the problem investigated. While the outcome of the first model will answer questions of how much gross-product is needed for a certain amount and for a certain structure of final demand, the second one will answer questions about effects created by the interrelations between different sectors or levels of hierarchy.

Structural change and large scale econometric models

Particularly in a period of fast structural change the link of econometric models to data of the past only can be misleading. In the case of transition-models of the former centrally planned economies towards capitalistic ones this restriction will lead to abandon this method at all because the behavior of the economic subjects will differ from the former behavior. In many cases not even the subjects will remain the same. Its identity may be changed, too. While the relationships in smaller models may be corrected by a-priori knowledge and by additional information about the new structure, and these relations may be tested against opposing hypotheses, for large scale models the demand for information may become that high that it cannot be managed any longer. If there is just a small amount of "soft" variables and relationships in a model some intuitive feeling about their interaction can be generated. In the case of a large scale model (particularly a nonlinear one) usually many "soft" variables and relationships have to be included. Thus the model builder will get lost in trying to understand the underlying dynamics of the large model.

Information pollution and data graveyards

Very large scale models may create information pollution. They could produce so many data that human beings get lost if they try to end up with any possible comprehensive view. Such large models can be of use only if there is some information compression feature available. Information compression could be done by computers or by human beings, or by an interaction of both. Later on under the headings of "deputy observer" and "automated impact analysis" the author will come back to possible solutions on the machine level. Besides that the usual way will be to "duplicate" the models structure by the organizational structure of the project team. For each sector or level of the model there must be some responsibility by individuals or small groups. In plenary sessions the outcome of the model with respect to a certain scenario in a first step can be evaluated by each of the subgroups and brought together by means of discussions afterwards. In such a case a project management is needed which is fully aware of its own necessity. Otherwise the person/s who has/ve to formulate the final report will have the full burden of drawing (or more correctly spoken "by inventing") conclusions. She/he will be overtaxed by such a responsibility.

Possible instabilities

Recent findings in chaotic systems research have shown that some classes of dynamical models may produce output which is affected by random elements although the equations are defined in a unique and deterministic way. That models amplify small changes in the initial conditions or parameters over the simulation period such that after some time (forecasting horizon) the random part of the output predicted will dominate the systematic part. It is evident that such models should not be used for forecasts for a period longer than the forecasting horizon.[12] Not only large models may show instability or chaos. But a particular difficulty arises with large scale models. The proof of chaotic or unstable behavior can be much more complicated for reasons of size and of additional numerical instabilities. But if one cannot proof the stability of the model one cannot be sure about the reliability of the forecasts.

6 Two unsolved problems

In generalizing his experience in model-building the author finds two complexes of un- or ill-solved problems. The first one is related to qualitative change, the second is the answer to the simple question "why?" in a models context.

How to handle qualitative change

It seems evident that there are difficulties to express qualitative change by mathematical models. The reason is that quantitative variables have a fixed qualitative meaning during simulation and that the identity of a variable has to be well defined a-priori. Otherwise we could not relate this variable to some observations of real life. Nevertheless we experience qualitative changes, not only where economic systems in Eastern countries are changed but in capitalistic economies as well (see the Fordism and Post-Fordism debate in France and FRG by the so called "school of regulation"). In the authors opinion there are several possibilities how to reflect qualitative change:

- First, one could redefine qualitative change by quantities (e.g. to measure qualitative change in evolutionary processes by a quantity like entropy, information or degree of complexity. Nobel prize winner Prigogine has used this method to describe the transition from single molecules to chains of molecules).

- Second, in the case where one variable changes its qualitative content one could include two variables of different meaning from the beginning of the simulation, but with variable weights[13] (in the limit before the qualitative change the weights could be 0 and 1, 1 and 0 afterwards).

- Third, one could construct a "deputy observer" of the model, i. e. a computer program which observes the models behavior. It creates or shows/hides verbal statements related to particular configurations of and/or quantitative relations between variables. The statements will be switched on and off depending on the occurrence of well defined events and shall signal if a new quality in the model occurs. This Deputy Observer could be constructed on the basis of mathematical logic, more advanced techniques could apply artificial intelligence methods for automated reasoning.[14]

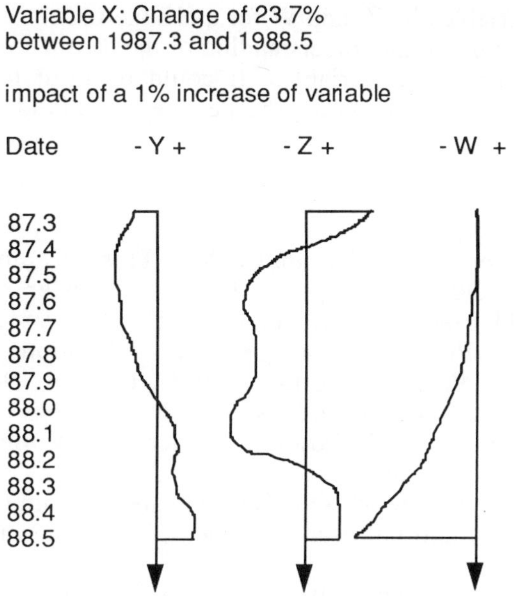

Figure 18.6 Hypothetical result of automated impact analysis

Automated impact analysis

Also if a nonlinear and dynamic simulation model is relatively simple the question: "Why does variable X change by 37.5 per cent between January 1978 and January 1979?" is by no means easily to answer although every piece of information needed is available in principle. In everyday economics the answer to the above question is given by some hint to changes in some exogenous variables, e. g. "because exports were expanded by 5 per cent". But this is no general and complete answer to the question. To handle and to operationalize this task the author thinks of a method where the impact of any of the variables can be visualized not only for one point in time but for a time interval. Theoretically it could be managed by using partial derivatives w. r. t. the exogenous and endogenous variables as well as w. r. t. time on the following basis:

$$dy_i = \left(\sum_j \frac{\partial f_i}{\partial y_j} \frac{\partial y_j}{\partial t} + \sum_k \frac{\partial f_i}{\partial x_k} \frac{\partial x_k}{\partial t} + \frac{\partial f_i}{\partial t} \right) dt; \quad i, j = 1, 2, \ldots n; \; k = 1, 2, \ldots m$$

The y_i represent endogenous variables, the x_k the exogenous ones, the f_i represent the right hand sides of the differential equations. The result could be a graph where the strength and direction of the impact of all variables is shown over the time interval under investigation (see figure 6 for a possible result with three variables Y, Z, and W for a change in X). Such a program could be useful to improve the understanding of the causal structure and the effects of some changes in variables. It could be used for any solved continuous dynamic simulation model irrespective of the field of application.

Notes

1. Jay W. Forrester, World Dynamics, MIT Press, Cambridge, Mass., London 1971; Other Writings by Jay W. Forrester are: Industrial Dynamics, MIT Press, Cambridge, Mass., London 1961; Principles of Systems, MIT Press, Cambridge, Mass., London 1968; Urban Dynamics, MIT Press, Cambridge, Mass., London 1968; Collected Papers, MIT Press, Cambridge, Mass., London 1975
2. Meadows, D. H., D. L. Meadows, J. Randers, W. W. Behrens III, The Limits to Growth, London 1972
3. Fleissner, P., Das Österreichische Gesundheitswesen im ökonomischen, demographischen und politischen Kontext, Vandenhoeck & Ruprecht, Göttingen 1977
4. Naschold, F. (Ed), Systemanalyse des Gesundheitswesens in Österreich, Eine Studie über Entstehung und Bewältigung von Krankheit im entwickelten Kapitalismus (Systems Analysis of the Austrian Health Care System - a Study about the Causes of and the Coping with Ill/Health under the Developed Capitalism), durchgeführt im Auftrag des Bundeskanzleramts am Institut für Höhere Studien, Fachverlag für Wirtschaft und Technik, Wien 1975
5. For Austria see: Fleissner, P., Wirtschaftsprognosen zwischen Orakel, Politik und Wissenschaft (Economic Forecasts between Oracle, Politics, and Science), in: Wirtschaftspolitische Blätter, Wien September/Oktober 1980, 37 - 49
6. Fleissner, P. (Ed), Systems Approach to Apppropriate Technology Transfer, Proceedings of the IFAC Symposium, Vienna, Austria, 21-23 March1983, Pergamon, Oxford, New York 1983; Fleissner, P. (Ed), Technologie und Arbeitswelt in Österreich (Technology and The World of Labour in Austria), Verlag des ÖGB, four volumes, Wien 1987; Fleissner, P. et al., Industrieroboter in Österreich, Verlag des Österreichischen Gewerkschaftsbundes, Wien 1984
7. Bundesministerium für Wissenschaft und Forschung (Ed), Mikroelektronik - Anwendungen, Verbreitung und Auswirkungen am Beispiel Österreichs (Microelectronics - Applications, Diffusion and Effects in Austria), Springer-Verlag, Wien 1981
8. Bruckmann, G. und P. Fleissner (Eds), Am Steuerrad der Wirtschaft (Controlling the Economy), Ein ökonomisch-kybernetisches Modell für Österreich, Wien New York 1989
9. Fleissner, P. and U. Ludwig (Eds), Ostdeutsche Wirtschaft im Umbruch (East-German Economy in Transition) - Computersimulation mit einem

systemdynamischen Modell, Braunschweig/Wiesbaden 1992, coauthors E. Biebler, H. Böhme, D. Dathe, G. Krüger, P. Nick und W. Oelschlägel

10. Fleissner, P., St. Adamec, St. Condik, F. Hajnovic, J. Haluska, K. Leutgöb, J. Meszaros, M. Olexa, J. Orsagova and M. Pohl, Ein Systemdynamikmodell für die Slowakei (A System Dynamics Modell for Slovakia) - Interim Report zu einem vom Bundesministerium für Wissenschaft und Forschung finanzierten Forschungsprojekt, Wien Juni 1992, partly in English
11. The predecessor of STELLA was PROFESSIONAL DYNAMO PLUS, a simulation language on mainframes and MS-DOS personal computers (see, e. g. George P. Richardson and Alexander L. Pugh III, Introduction to System Dynamics Modeling with Dynamo, MIT Press, Cambridge, Mass., London 1981). For a detailed description of STELLA, a user friendly simulation software for the Apple Macintosh, see Barry Richmond, Steve Peterson, Peter Vescuso, An Academic User's Guide to STELLA, Lyme 1987
12. For a quantitative measure of the forecasting horizon for nonlinear input-output models see: Fleissner, P., Dynamic Leontief Models on the Test Bed, in: Structural Change and Economic Dynamics, vol. 1, no. 2, 1990, 321-357, 356
13. To describe the transition to a CIM-economy we have constructed a model with two sectors, a traditional one and a CIM sector. Over the years the new technology spreads all over the economy. Finally a saturation level of CIM-technologies is reached (see Fleissner, P., and W. Polt, Switching from Fordism to CIM: A System Dynamics Model, in: Ayres, R. U., R. Dobrinsky, W. Haywood, K. Uno and E. Zuscovitch (Eds), Computer Integrated Manufacturing, Chapman & Hall, International Institute for Applied Systems Analysis, London etc. 1992, Volume IV, 391-406)
14. Maybe a program like NueX by Charles River Analytics, Boston, USA, can serve this purpose. It is written on top of HYPERCARD and supports simple production rules in an expert system module.

19 The genetic equipment of scientific and innovation activity

Ladislav Tondl

1 Introductory remarks

Even the most inexperienced builder will realize that if he wants to build a stone house he will need stone, and if he intends erecting a brick house he will have to arrange for bricks to be ordered, purchased and manufactured somewhere. It goes without saying that this does not mean that a new structure is made up solely of a heap of stones or bricks, that it will simply be built on its own, without any higher design whatsoever, without following specific objectives, programmes or projects, and especially without necessary knowledge, experiences, skills, without vital tools and technical instruments. In a sense, all these material and intellectual prerequisites tend to predetermine the fate of a new artefact. Indeed, they invariably encompass certain factors, which tend to affect - to a smaller or greater extent - the development and utilization of the given artefact, the quality of its functions and the time horizon of an efficient application of such functions, eventual risks connected therewith, the overall period of full application and hence also the moment of inevitable or expedient termination of such functions. In a sense, the origin, beginnings, anticipated development or anticipated behaviour patterns of these man-made artefacts whose interplay is sometimes figuratively described as a "second nature", displays a number of analogies with the genesis, behaviour, development and extinction of objects of "live nature". One can, therefore, consider some authors' claims referring to "biological metaphors" as being justified (this term was used by H. Brooks, author of numerous studies on development trends in science and technology and head of teams of authors of major OECD studies on science and technology policy), in terms of "genetic equipment", a set of "genetic attributes", "genetic code" (in Czechoslovakia, it was J. Vlçek who singled out the significance of this term in systems engineering), to the emergence and demise of scientific concepts and paradigms, to various generations of certain technical systems etc. It would therefore be possible to highlight some other forms of biological metaphors; with many artefacts, it is plausible to distinguish stages, which are analogous to normal and pathological states, and it is likewise feasible to single out situations corresponding to that which comes under the heading deviations, mutations or any similar divergences in biological systems.

Biological metaphors relating to human artefacts, both purely intellectual ones and those of a technical nature, are known to feature various dimensions out of which the following should be considered as significant:

- a certain information charge, which is always tied up with a capacity involving memory and time intervals;
- a certain entropic, or non-entropic process, i.e. a process of changes marked by entropic nature (or level of arrangement), both processes of desirable changes, i.e. the creation of artefacts with a reduced level of entropy, as well as processes of undesirable changes, i.e. increase in entropy;
- causal structures of artefacts created as open systems, which exercise certain impacts on their environments, while coming under outside influences.

One can therefore raise objections to the effect that the above-mentioned characteristics of biological metaphors, in the form given here, tend to display a predominantly physical nature. However, the application of these salient features should not be omitted even in those fields of rational activities designed to pave the ground for changes in science and technology. Even in this particular sphere the following rule is valid: neither desirable nor scheduled changes can ever occur without appropriate prerequisites, they cannot be effected without the presence of a large-scale set of instruments, which can be collectively described as genetic equipment. Or to put it in other words, no "information perpetuum mobile" can, nor does exist in this domain.

2 Genetic equipment in technical and innovation policy

Within the technical and techno-economic sphere, one can frequently encounter a grossly simplified perception of the complex nature of such prerequisites, which come under the collective heading "genetic equipment". The core of this is seen to inhere in material or energy sources and existing production capacities. Having vast resources of kaolin and several manufacturing plants boasting fine tradition in this field at our disposal, we can therefore produce and export good-quality porcelain. Furthermore, these products are greatly appreciated and sell very well abroad. This kind of simplified notion of the structure of genetic equipment may easily be transferred to other areas as well. If some elements of such a simplistically perceived structure are found missing they may be added, created anew or imported. There are some well-documented cases from the past few decades in our country exemplifying such a procedure. Mining capacities were simply created on the site of the existing vast deposits of brown coal, and coal-burning power stations were added later. The existing large-scale heavy engineering capacities in North Moravia were expanded by other facilities with identical functions, the supplies of (a very substandard) iron ore had been provided and such a power-intensive production was then considerably expanded. It is also evident that the afore-mentioned, oversimplified interpretation of genetic equipment could only be associated with those growth concepts, which have since been dubbed "incremental" innovations. It should also be noted that this particular concept does not consider what have been termed substantial dimensions of emerging artefacts, i.e. information aspects, memory characteristics, time horizons, possible or anticipated changes in the level of entropy and especially the complex causal structure, including the large-scale spectrum of adverse ecological, cultural and moral impacts on the physical and human environment.

Neither in the realm of technical and innovation policy is it possible to reduce what has been called "genetic equipment" and what actually amounts to a very intricate complex of a priori prerequisites of any changes, desirable development patterns, required or scheduled changes solely to material resources and to what - in the spirit of the economic theories of the past centuries - used to be termed "forces of production". (Such theories can be neither justified nor improved by claims that science too is "an immediate productive force", as maintained by reformers of these theories).

A characteristic trait of the traditional scientific approach to analysis, comprehension and explanation of all the changes, including those resulting from man's rational activities and those instrumental in the creation of new artefacts, is the gist of what the question "why" really encapsulates. To put such a query in the given contexts usually means calling for a scientific explanation, a specification of causes and causal chains. It is quite well known that a specification of such demands has been instrumental in shaping concepts of elementary models of scientific explanation proceeding from the knowledge of deterministic or probability dependences and empirically characterized initial states. To put it in other words, in order to explain a specific phenomenon, a specific event on the basis of such a conception, the following tasks should be undertaken:

- to specify the initial state, point of departure, or what can be termed as a complex of causes,

- to specify relevant dependences, e.g. in the form of what are known as scientific laws, hypotheses, rules valid in the realm of phenomena or processes among which one may include the initial state (this, of course, corresponding to the concept of the rules in the field of expert systems, or generally in artificial intelligence, most frequently in the following form: "if A occurs under the circumstances of B then C exists with certainty or with a certain measure of probability").

- to determine a system of rules according to which data covering the two given circumstances can be used to derive information on this particular phenomenon or event, which, in turn, is the subject of explanation, i.e. a set of inferential rules.

Obviously, the afore-mentioned pattern of scientific explanation corresponds to the image of the world, which constitutes a certain generalization of traditional ideas and notions proceeding the principles of physical processes established by Descartes and Newton. These ideas may be supplemented with Laplace's notion of the demon or "absolute intelligence". If we had a complete and exhausting description of the initial state, as well as thorough knowledge of all the dependences at our disposal, we would be in a position to explain virtually anything.

Within the context of this study, which is concerned primarily with delineating genetic equipment of technical and innovation activities, we will try to demonstrate that on-going trends in institutional or organizational patterns are not capable of ensuring the implementation of sweeping and desirable changes by themselves, simply because genetic equipment invariably enters as a heritage of the past, as a set of genetic information, characterized by a specific amount of inertia, memory and time horizons. Having said that, we naturally do not claim to contradict the key role played by changes in existing institutional or organizational patterns. Reverting to the afore-mentioned question "why", or rather to the query concerning explanation of the genesis of technical artefacts, the sum total of what has come to be described the specification of the initial stage and relevant dependence (a set of data pertaining to these fields is traditionally referred to as "explanans" in relation to the data spelling out the object, phenomenon or event which is to be explained, i.e. "explanandum"), then the specification of the initial stage and relevant dependences tends to be much broader than in the case of patterns involving the explanation of natural phenomena or processes. A technical change, a new technical artefact always results from a very intricate interaction involving diverse circumstances, which should, in no case whatsoever, be reduced solely to what some economic theories of the past century termed material resources and forces of production. As far as major circumstances co-participating in primary decision-making on the possible or contemplated technical change are concerned, the following should be mentioned in particular:

- an awareness of the possibility of changing the existing status quo, the required change, the need of drafting a new technical solution, determining an objective of such a solution etc., hence specifying a set of circumstances and conditions of a subjective nature,

- a set of findings on resources, instruments and methods for the implementation of contemplated technical solutions, or the production of technical artefacts, as well as the knowledge of technical artefacts, including the knowledge of the pros and cons of individual feasible variants, and their potential or anticipated impacts in time and space,

- knowledge of the feasibility conditions and possibilities of individual variants, of the functional, temporal and other horizons of such variants,

- a system of values or preferences, which co-participate in decision-making, assessment and selection of individual variants of solutions.

With emphasis on the fact that the given circumstances, factors or knowledge and value-oriented impacts co-participate in the genesis of technical solutions. In the creation of technical artefacts, however, they should be placed into context with the awareness that such circumstances or influences invariably result from long-term development, that they establish themselves as a gradually created memory factor, which continues operating in new situations, under newly arising social and political conditions. That is also why a mere change of conditions or a change in the existing organizational and institutional structures usually does not bring about a far-reaching change in the content and function of the entire sphere of problems involved. That is why genetic equipment, which was created under previous historic circumstances, is known to have been applied or at least - to have been applied over a specific period of time - under these particular circumstances.

These predominantly general reflections on the role of factors affecting the technical and economic bahaviour and decision-making, or rather the survival of genetic equipment in the given areas, may be transposed onto an analysis of this particular situation, which in my own country should be attributed to the long-term impact of the former regime of monopoly power, a system of directive decisions, of "delegating" authorities and competences from the highest echelons of monopoly power. Asked about the repercussions of the long-term impact of such principles, organizational and institutional structures, we often single out a heavily polluted living environment caused by an erroneous technical and investment policy, excessive material and energy consumption of numerous manufacturing sectors, a technological delay spanning, in some spheres, several decades, substandard performance of services and other fields of infrastructure, an adversely affected health situation, exacerbated particularly by the prevailing environmental situation (which should, however, be perceived not only within the context of a polluted physical environment but also as the devastation of interhuman relations, moral values, personal and social responsibility).

The downfall of the regime of monopoly power, coupled with the emergence of an open, pluralistic and democratic society, the takeoff of a market economy, the legislative provision and safeguarding of fundamental human rights as well as numerous other changes are bound to pave the way for uprooting the main causes of the above-mentioned legacy of the past, but they can hardly aspire to remove, either immediately or over a short period of time, all the negative repercussions of this heritage. Not only has virtually zero progress been made over the past two years or more in improving the living environment in Czechoslovakia's most polluted regions, in reducing power and raw materials consumption of many manufacturing sectors and in correcting the unfavourable health situation of the nation. However, compared with the recent past, there is a considerable difference nowadays, namely that these negative impacts can be, and duly are, openly reported on, and that they are no longer concealed and obscured by an official ideology. A much less pronounced feature of the present-day life in Czechoslovakia is the survival or perpetuation of the past despicable patterns of human behaviour, way of thinking and reasoning, and values determining people's decision-making. A justified claim can be made that a technological and economic gap along with a delay in the level and quality of artefacts is matched by a delay in the spiritual, cultural and moral sphere, in terms of people's knowledge, skills and value-related structures. In practical terms, this means that citizens entering the newly emerging socio-political and economic life inevitably carry their own genetic equipment with them, which has been created by the past two generations and over a period of several decades. (In actual fact, this time horizon could cover not only the forty years of totalitarian communist rule but also the period of German Nazi occupation during World War II.) Given below are just a few examples documenting the ingrained habits, ideas or value-related attitudes, which co-shape the genetic equipment in the given sense of the term.

These systems of monopolised power were based on those principles according to which any initiative capable of stimulating major technical or socio-economic changes could and should come solely "from above", i.e. from the top of the system of monopoly power, usually accompanied by a measure of high-handed behaviour, infallibility and insistence on the absolute necessity of employing nothing but the one solution chosen by the top brass. So far, many people, even though realizing their newly gained freedom in decision-making, still tend to act according to their deep-rooted behavioural patterns, expecting that major changes would come from the top or even demand that such changes should come from such quarters, as demonstrated by several industrial strikes or strike alerts in Czechoslovakia lately. The reverse side of this coin is the emergence of a seemingly opposite situation whereby certain groups of people sharing common interests display similar elements of high-handed and infallible behaviour so typical of the centres of monopoly political power in the past.

The previous system was based on the egalitarian notion of considerable levelling-off in remuneration, notably in salaries and wages (the only exception being, of course, the top echelons of the monopoly power centres), on what can be termed as the "dictatorship of the mediocre". Actual competitiveness in terms of knowledge, abilities and skills was replaced by a formal farce pretending to be competitive, whether this was called "socialist emulation", "socialist labour brigades" etc. In actual fact, that competitive edge, which may be dubbed "competitive capacity", has all but disappeared. That is also why, particularly during the early days of the functioning of a market economy, the competitive capacity, an ability manifesting itself in the quality of products, in an ability to take advantage of a longer time perspective, in the principles of modern marketing or some analogous means and methods, has been taken over by subjects who - entering this environment - engage in procedures which are on the borderline of legality and morals and which, in some instances, overstep those boundaries. It is appropriate to add in this context that the process of creating such a competitive capacity is, to all intents and purposes, a long-term affair, depending on the ability of individuals to accumulate considerable knowledge and also rich experiences. This holds particularly true of what are called entrepreneurial activities and entrepreneurial initiatives.

3 The importance of genetic equipment in science

The significance of these factors or sources of scientific initiatives has been discussed in a large number of books, particularly on the history and sociology of science. The first group of such works includes studies on the origin and development of individual scientific schools, on the continuity of scientific generations, on the impact of outstanding personalities on their successful pupils and the like. It is a well-known fact that, as a rule, this is a long-term impact, and not even the adverse effect of the circumstances prevailing in Czechoslovakia over the past decade could completely destroy the fine traditions of some of its scientific schools, e.g. in chemistry, mathematics and other natural sciences and partly in technical disciplines. Yet another group of works covers sociological studies analyzing the social contexts of creative intellectual activities, which examine social and intellectual relationships within research teams, studies on the sociology of learning etc. The earth shattering studies by Merton, and the works by Ben-David on the genesis of scientific achievements which were awarded the Nobel Prize etc. should also be mentioned. Other publications include analyses of the impact of negative changes of leading personalities, namely if a truly creative and stimulating personality is replaced by somebody who is more or less a "scientific clerk". In actual fact, the history of science in Czechoslovakia over the past few decades has caused untold damage whose repair will be a long-term process indeed. It is a history marked by numerous insensitive administrative interventions, like a brain drain during waves of emigration of competent scientists, and the application of the principles the author of this study has termed "intellectual Darwinism upside down". Up till now such phenomena have been recorded in Czechoslovakia solely through scientometric methods, notably in terms of the country's slide down the ladder in the world's scientific and intellectual league. A more comprehensive historical and sociological analysis of these phenomena is yet to be carried out just as an analysis is needed to probe similar phenomena in various fields of Czechoslovakia's cultural scene.

An attempt will now be made to highlight several groups of factors affecting the success, efficiency and productivity of scientific activities, or rather to show the long-term impact of various forms of work, value-related attitudes and established social conditions, and thus sets of circumstances, which are passed on into further work and which would therefore be taken into consideration in upcoming research programmes and projects. As a result, this involves a certain type of genetical equipment, which finds its expression in the new programmes and projects. Such groups of factors include in particular:

- existing intellectual clusters, or rather contemporary or anticipated development patterns of such clusters,

- well-functioning intellectual networks, link-ups to foreign research centres, including forms of cooperation with these centres,

- social and value-related "background" to research activities, response to and evaluation of the results of such activities, as well as reception of these results,

- a demanding environment which may be described as a "competitive environment", one that appreciates major results and is highly critical of substandard activities or any negative phenomena in scientific life.

A successful scientific team, a major scientific school or, generally speaking, what has lately come to be known as an **intellectual cluster**, especially in a research centre, or in a university centre, usually comes as a product of long-term impact of highly stimulating scientific personalities, of an intellectually demanding and critical climate as well as an environment, which is in a position to appreciate fully, grasp and interpret research results that are known to constitute an appreciable step forward correctly. That is also why such clusters can hardly originate by virtue of a decree or resolution issued by an administrative body, an institutional, organizational change or an administrative measure usually presented as a "reform", "reshuffle" etc. On the contrary, it is a well-known fact, fully confirmed by Czechoslovakia's experiences over the past few decades, that ill-considered administrative measures (usually motivated by political, ideological or other political ambitions) have been corresponsible for the demise or destruction of certain, formerly successful and acknowledged intellectual clusters. While totalitarian systems were marked by tendencies towards uniformity in all walks of social life (shirts and scarves of identical colour worn by children and youths, an amorphous crowd spending national holidays by paying tribute to the country's leaders or representatives as they used to be called in the past few decades), an equally typical trend towards uniformity coupled with the dictatorship of mediocrity manifested itself in the intellectual sphere. In practical life, this meant that an eminent scientist could easily find himself in trouble as long as he stood out too much, or find that he was paid the same salary as his mediocre colleague ranked in the same professional category and endowed with the same title, the latter being so rewarded solely for his "faithful service" or at least for his conformity with monopoly power. The above-mentioned features aptly reflected the cult or at least over-emphasis on administrative and organizational changes. While looking back on the numerous reforms in Czechoslovakia's educational system, including university education, reforms in the research planning system, in different ministries and top-ranking commissions, charged to "manage" research and development, one may easily realize that the cult of administrative changes has indeed grown to be part and parcel of that heritage or genetic capital we have now inherited after long decades of the past administrative system and the past value structures. That is also why many people are still mostly preoccupied with issues such as who will be subordinated to whom, what kind of managerial system our research centres will eventually adopt, while the term management is still bound up with many traits typical of the old interpretation of the term, i.e. strict sense of subordination, high rate of commands or directives coming down from the top as well as one's necessity of filing a great number of reports, statistics etc. to the top. Furthermore, many boards of management, acting according to well-established traditions or valid rules responsible for the management of large research centres, still tend to express themselves with a great deal of self-confidence even on issues they are utterly unqualified to discuss. Naturally, this claim is not intended to play down the significance of those legislative changes, which have been instrumental in abolishing the monopoly of certain organizational structures, introducing elements of autonomy and pluralism in the scientific and academic communities in Czechoslovakia. These measures have, quite justifiably, been designed to weed out the normative and organizational sources of countless negative phenomena. At the same time, one should certainly not overestimate the impact and importance of such measures. That is also why formal, administrative and normative changes by themselves, without changes in the contents, programmes and objectives of research activities can neither restore nor revitalize intellectual clusters, nor can they do much to safeguard their permanent reproduction with an influx of young talent and especially to enhance the overall attractiveness of such activities.

As a matter of fact, outstanding research results are accomplished not only by an isolated genius, they often come as a result of well-functioning information networks, mutual stimulation, mutual criticism and generally excellent forms of communication both within research teams or large institutes and among various experts, reseach centres etc. Therefore, priority is often given to the significance and role of what science and innovation policy programmes of the end of this century call a **system of intellectual networks**. This particular aspect has lately been highlighted by the integrating tendencies in the economy, technology and culture, which are part of the process of integrating Europe, by the downfall of various barriers, symbolized by the Berlin Wall. The existing spectrum of forms and possibilities of intellectual networks is naturally very rich, encompassing congresses, conferences, different joint projects, direct professional contacts or information exchanges, including so called invisible colleges, exchanges of hitherto unpublished texts or studies in print etc. One particularly successful format in recent years has been the workshop, a working meeting bringing smaller groups of experts from various research centres together for discussions where new problems and issues are identified and posed, and wholehearted efforts are made to stimulate what amounts to be the most important component of the cognitive process, i.e. that insatiable desire for learning, curiosity, a yearning for what may ultimately turn our to be an analogy to aesthetic experience (since new findings can often thrill the researcher with their beauty or elegance).

A well-functioning system of intellectual networks can hardly be established by virtue of an administrative decree, organizational or institutional changes; instead it is introduced gradually, providing for the establishment of a climate conducive to mutual confidence, mutual understanding and mutual respect. The long-lasting isolation of Czechoslovak research centres in the past, caused by the division of Europe, by the political, economic and ideological boundaries but also by a lack of financial funds, certainly did nothing to encourage the creation of well-functioning intellectual networks. The only exceptions to the proverbial rule were individual researchers and scientists who - more often than not - had to grapple with the misunderstanding or sometimes outright disfavour on the part of the official science administrators. Indeed, the former communist authorities would formally proclaim and endorse support for certain integrating tendencies and projects, but this would primarily concern trends within the Soviet bloc countries and would often involve backing for cooperation with research institutes of dubious quality. Viewed in these contexts, one can hardly escape the conclusion that efforts to overcome the miserable heritage of the past decades in Czechoslovakia will be neither easy nor short-term. Moreover, nowadays there are very few younger and middle generation scientists and researchers in Czechoslovakia who have spent several years or at least a few months on study stays at leading universities or research institutes abroad. Seen in this light, Czechoslovakia is facing a dire lack of that genetic equipment, which is so vital for an efficient functioning of intellectual networks.

Similarly, the term **"social and cultural background"** of research activities refers to a complex of diverse phenomena and factors. Under such circumstances it is usual to revert to the spirit and tradition of outstanding ancient universities or research centres, noted for what used to be the overriding meaning of philosophy, i.e. respect for wisdom. This should also include what may be termed as a healthy moral climate, as an awareness of responsibility and horizons of this responsibility. It is next to impossible to ignore that this particular social, cultural and moral climate has been severly impaired in Czechoslovakia over the past few decades, mostly due to several major factors, associated with the former regime of monopoly power, with the phenomena we have collectively described as "intellectual Darwinism upside down" or "the dictatorship of mediocrity", with the enforcement of ideological and political conformism. It is likewise essential to realize that the elimination of the major root causes of these problems does not automatically mean the abolition of their consequences, namely in the field of work habits, attitudes to assigned tasks, programmes and projects. Up till now many people in Czechoslovakia have not yet managed to change their approach to research programmes to any great extent, even though the latter are no longer "assigned from above", and even though the former are now absolutely free to participate in the drafting of the concept of such projects.

Such a climate of scientific and research activities is also linked with what may be called a **competitive environment**. First and foremost, a competitive environment in research and development represents a highly demanding and critical climate involving the activities of individual research teams. Although the cult of "criticism and self-criticism" used to figure prominetly among the key features of the official ideology in Czechoslovakia's recent past, an atmosphere of criticism never prevailed in the scientific community. One of the reasons for this was the justifiable apprehension that honest criticism might be abused for administrative intervention. Many scientists retained vivid memories of the ideological campaigns conducted during the 1950s, campaigns that were accompanied by numerous administrative measures and a great deal of persecution, especially in the field of humanities. Needless to add, such concerns took quite a long time to allay and overcome, and a sense of false professional cohesion and togetherness developed as a result. Seen in this context, the introduction of grant systems should be perceived as a major step towards promoting a competitive environment in Czechoslovakia's scientific life.

4 Some final recommendations

By emphasizing the key role of the above-mentioned factors and long-term conditions operating in the given spheres in terms of scientific and innovation activities, we certainly do not intend suggesting that we should in any way succumb or even subordinate ourselves to the impact of these factors. No doubts should be harboured whatsoever about the need of effecting a far-reaching transformation in these sectors, a significant role is being played here by the objectives, concepts, anticipated functions and required standards as well as the concept of vital organizational, institutional and normative prerequisites that would facilitate a gradual approximation to the stipulated goals. Such objectives inevitably include the promotion or restoration of what have earlier been described as "intellectual clusters", the creation of efficient intellectual networks and the stimulation of a suitable social and cultural background. While aspiring to tackle these particular issues, we can hardly ignore the initial state or point of departure for desirable transformations, including a critical analysis of the available genetic capital that has to be taken into account in such projects. The fact that within the programmes and projects, which can ensure suitable forms of approximation towards stipulated target structures, many negative features of the past heritage must be taken into consideration and should not be explained away by saying that we are simply willing to reconcile ourselves, or put up with this legacy. Even within the intellectual community such a heritage is still known to feature a reluctance to conduct a dialogue (since insistence on conducting a monologue, usually a monologue dictated from above, figured among the salient features characterizing monopoly power and its ideology), disrespect for a plurality of opinion, attitudes and values, and especially an exaggerated feeling of one's own self-importance now displayed by many people who for long years condoned or silently tolerated the authoritarianism enforced in virtually all walks of life by the former social and political system in Czechoslovakia.

20 Scientific occupations in Czechoslovakia

Eva Frydová and Adolf Filácek

In the 1960s, approximately a quarter of a century ago, research was carried out on the prestige of scientific occupations and on those occupations requiring a tertiary qualification within the framework of research into the standing and role of research workers in Czechoslovakia. This research into public opinion put Czechoslovakia as the 60th country in the world in comparison with research carried out internationally during the 1950s and 1960s.

The results of this research were auspicious for those professions which require a tertiary qualification. Within the boundaries of thirty highly-rated professions, six of these were placed in the first ten. What is more significant is the fact that five of them took the first five places in the order of the professions' general social prestige. Among these was the occupation of scientific worker, which came fourth.

The researchers interpreted their results as being connected, among other things, with our country's economic situation at that time, and the bodies which governed according to the declared concept of techno-scientific revolution, which was able to afford the publically marked illusions about the unlimited possibilities for science to resolve all manner of economic and social problems. The representatives of society's so-called scientific administration then created the fiction of a rational society, in which all determining social processes are scientifically described and regulated. Scientific knowledge was made to play the role of a stimulant, automatically capable of removing the majority of its own deficiencies.

This simplified view of science's role intensified in the 1970s and 1980s. One of the consequences of this was that economic failures as well as deficiencies in administration and in the way social life was organized were often attributed to the low level of research and the inconsistent application of scientific knowledge in practice. The immense long-term problems of the national economy in the 1980s were considered above all to be a result of the very slow rate of acquiring new techno-scientific knowledge in manufacturing and consumer goods, and to an insufficient extent.

The unfit nature of the decision-making areas of the national economy's administration was, however, reflected in the administration and organization of science as well. The extensive network of research bases did not provide sufficient results. As a result of not providing these bases with enough apparatus and necessary materials, the lack of contact with scientific work worldwide and the inadequate preparation of the up and coming scientific generation, some branches of science gradually fell behind the world standard. The field of science as a whole was under the influence of enforced Marxist ideology, which slowed down the progress of different scientific branches, particularly in the social sciences. Applied research did not produce enough applicable results, and in manufacturing the desired utilization of new knowledge did not come about. The myth of a close communication between research and production slowly melted away. Science did not fulfil the social function which the leading state organs envisaged and demanded of it.

The above circumstances (and others) relating to the mission and role of science in society led us to think about how this unfavorable situation in science had produced a repulsion in peoples' consciousness, as it influenced the prestige of scientific (research) occupations, as well as the prestige of science as a whole.

The concept of our research into the prestige of scientific (research) occupations, carried out in November 1990, took the possibility that the public would become skeptical as to the scope, significance and role of science in our present society into account, and that this would also be expressed in the appraisal of prestige of scientific and research occupations, all as a result of the mentioned distortion in the development of science and research.

The research into the prestige of scientific (research) occupations in our society was carried out as part of a research project entitled "Social Differences and the Prestige of Professions" (from now on to be referred to only as Prestige '90), which the Institute of the Theory and History of Science of Czechoslovakia's Academy of Sciences and the Institute of Sociology Czechoslovak Academy of Sciences have worked together to produce. A representative sample of 1400 respondents was taken from both the Czech and Slovak republics for this research into public opinion.

Its main aim was to discover how the present public values scientific and research professions, what place they would give them in relation to other important professions, and how they rated the significance and role of science overall in the present period of changes in social structure and radical reforms.

To assess profession prestige, 50 professions were chosen for the research, representing the main professional groups in our society. In addition to this basic group of professions, the respondents were presented with a separate group of 15 occupations requiring a high level of qualification, which put further emphasis on scientific and research professions, itemized according to their specialist orientation.

The respondents were asked by those conducting the research (both groups were done separately) to rate the professions in accordance with what prestige they gave each one. The meaning of the word prestige was defined in the question "How do you personally regard people from the given profession?". The profession was given points on an untraditionally wide scale of 1-100, and the respondent was then invited to choose the profession with the highest prestige and give it number 99, and then to give the lowest prestige profession number 1. Only then did he give points to the rest of the professions. This methodological procedure was used in order to ensure the maximum possible consistency and mutual comparability of the answers (utilizing the whole scale). At the same time, it was used to enable the prestige scores obtained to be standardized with respect to the internationally comparable 100-point scale.

The following Table 1 shows the placement of scientific (research) professions in the basic group of 50 professions. Science or research are represented by 5 professions, which constitute different levels of research:

- scientist (basic research),
- associate professor, professor (basic research in tertiary education institutions - High Schools, Universities, etc.),
- research and development engineer (applied research),
- design engineer, designer (research in commerce) and,
- head of a research establishment (research administration).

Table 20.1

Evaluation of occupation prestige (basic group of 49 professions)

Ranking	Name of profession	Average prestige score	standard deviation	most frequent response		Evaluation over 90 points, frequency in %	Evaluation up to 10 points, frequency in %
				Mode	Frequency %		
1	Poly clinic doctor	80.4	19.0	90	12	47	2
2	Minister	75.6	25.6	99	14	48	4
3	Associate professor /Professor	74.5	20.2	80	12	31	2
4	Scientist	71.9	22.4	90	12	31	3
5	Company Director	70.9	21.7	80	12	28	3
6	Diplomat	68.7	25.7	90	12	33	5
7	Primary school teacher	66.4	20.5	80	10	18	2
8	Nurse	66.1	21.4	80	10	18	2
9	Miner	65.8	25.5	90	9	27	3
10	Head of research centre	65.5	24.2	80	11	21	5
11	Manager	65.1	24.3	80	14	22	4
12	R&D Engineer	64.3	22.1	70	11	17	3
13	Lawyer	62.8	22.6	60	10	16	3
14	Industrial economist	62.6	21.3	75	10	12	3
15	Operational engineer	61.4	19	70	11	9	3
16	Agronomist	59.6	19	60	12	8	3
17	Design engineer	58.4	20.1	70	11	7	3
18	Machine operator	57.7	20.2	60	10	9	3

Table 20.1 (continued)

19	Programmer	55.1	21	60	10	6	4
20	Farmer	55.1	24	60	9	11	5
21	Head of economic planning	54.9	23.7	70	10	6	7
22	Factory foreman	53.6	19.7	50	10	5	3
23	Writer, Composer	52.9	25.9	70	8	11	6
24	Concreter, Welder	49.8	23.1	50	9	7	5
25	Bricklayer	49.2	23.5	40	9	8	5
26	Journalist	48.9	22.5	50	11	5	7
27	Tractor driver	48.5	24.8	60	9	9	7
28	Professional sportsman	48.5	28.4	50	8	10	17
29	Car mechanic	48.4	20	50	11	3	3
30	Shop owner	48.4	21.4	50	12	5	5
31	Computer operator	48	20.7	40	11	3	3
32	Lorry driver	47	21.8	30	8	6	5
33	Policeman	46.5	25.9	50	9	7	14
34	Lathe operator	45.9	22.9	50	12	4	8
35	Priest	45.3	30.4	50	8	12	20
36	Adjuster	44.6	21.1	40	10	2	5
37	Tailor	44.2	20	40	12	3	5
38	Television repairer	43.6	18.9	45	11	1	5
39	Accountant	42.8	19.2	50	11	2	6
40	Bank clerk	41.4	21.4	50	10	4	10

Table 20.1 (continued)

41	Textile worker	41.3	21.9	40	10	3	8
42	Major	29.5	24	5	13	1	32
43	Secretary	29.3	18	30	12	0	18
44	Waiter	28.4	19.7	10	12	1	23
45	District official	27.4	19.5	5	12	0	27
46	Auxiliary worker	25	22.9	5	16	2	38
47	Cleaner	24	22.9	5	15	2	37
48	Political party secretary	21.7	23.9	1	20	2	49
49	Street vendor	18.7	17.6	5	17	0	44

From the table it can be seen that the prestige of research professions is noticeably high. Scientists came in fourth place with an average score of 71.9 points. Also significant is the fact that 12 % of the population surveyed gave 90 points to scientists, and 31 % of the population put their evaluation of them at over 90 points. We can verify the high prestige of scientists by further results obtained by a more in-depth statistical analysis. Statistically, the prestige of scientists does not depend on the respondents' age or education. This means that the evaluation of scientists is equally spread throughout the population. It can be concluded from this that this profession is respected by almost all social groups of the population. Considerable respect is also enjoyed by the profession of associate professor (professor), which came third in the professions' ranking on the strength of its 74.5 point average score, behind polyclinic doctors and ministers. At the same time it can also be stated that all 5 representatives of research came in the top 20 occupations. Out of these, the head of a research centre had a significant placing (10th), ahead of the research and development engineer (12th) and design engineer/designer (17th).

The ranking of the first 20 occupations provides interesting information. The first thing to note here is that, numerically, the so-called intellectual occupations outweigh the manual ones. Only 3 occupations deviate from this framework of intellectual professions: that of miner, machine operator and private farmer. In the intellectual professions, health is the area which is primarily represented, with doctor in the traditional first place. After that comes management and organization (minister 2nd, company director 5th, manager 11th, head of a research centre 10th), then science and research together with the professions already mentioned.

This ranking indicates which social areas the public considers to be important and socially beneficial. Foremost among these are the whole area of economics, including management and organization, health care, education and the related contribution of new knowledge and innovations in research. Peoples' attitude to merits is also reflected with the appraisal of these social areas. It can be stated that health, education and money are among the preferred merits.

Three professions, however, do not fit in at all with this statement; miners came in 9th place (65.8 points), machine operators 18th (57.7 points) and private farmers 20th (55.1 points). The prestige of miners has a tradition of many year's standing in Czechoslovakia. As an occupation, it particularly entered the public's consciousness in the 1950s (the boom in Stakhanovite movements and the estimation of hard manual work connected to this.). The continual level of risk involved in this occupation, its tradition and benefit to society, are the manifestly important criteria which give it its high rating. The public's reaction to the practical usefulness of these reflected in the placement of agronomist and private farmer. An analysis by factors of the professions' prestige rating (see further on for greater detail) showed that the respondents tended to stress more the actual aim of the occupation, while the dimensions of the private sector were conspicuously unapparent.

The complete ranking of 49 occupations provides some further points of interest. Two traditionally labouring professions came in the first half of the ranking: concrete-layers (welders) and bricklayers. Professional sportsmans also obtained a fairly high rating (28th). On the other hand, clerical occupations, light-industry workers and professional soldiers had a low prestige level. The placing of the last two is worthy of note: the last but one place was taken by political party secretaries, which can be attributed to the unfortunate consequences of the totalitarian period in our society, and to the fact that newly emerging political parties have not yet managed to build up the required respect. It is worth noting that street-traders were placed last (behind cleaning-women), in spite of the general and widespread image of their high earnings and spirit of private enterprise. It can be surmised that this is one of the consequences of people's surviving view of selling in the street as a "dishonest way of doing business".

In comparison with the research of Brenner, Hrouda/1966/ allows for a more detailed analysis of our results. A significant part of this work was research into occupation prestige, with a marked emphasis on the occupation of research workers and the group of occupations requiring a tertiary (high-school) qualification. The investigative research was carried out via a set of questions to a representative sample of 1400 people, on a state-wide basis. 30 occupations were selected for this survey (representing the main professional groups of society at that time) according to the following criteria: level of education, proportion of physical and mental activity, and the sector of the national economy. Seven of these occupations presupposed a high-school education (architect, engineer-design engineer, doctor, judge, secondary school teacher, research worker, teacher in high-school education). These occupations were chosen to represent the majority of workers in the national economy who had experienced tertiary education in one form or another.

Fifty occupations in all were taken in the ITHS Prestige '90 research.

These were selected so that individual occupations could reflect the contemporary structure of professions in society, and at the same time conform to the International Standard Classification of Occupations (ISCO) with the aim of providing survey results comparable with other countries.

In addition to this, the respondents were given a further 15 occupations to assess separately. These occupations required a high level of qualification and were heavily oriented towards scientific or research occupations. The relation of these occupations to the basic group of professions was ensured by using "anchor professions" -professions incorporated into both groups - and by using the same methodology to assess occupations.

A comparison of both surveys referred tothe Brenner, Hrouda/66 research and Prestige '90 calls for a brief reflection on some methodological approaches, particularly bearing on the criteria which provided the basis for respondents to assess the occupations.

In the Brenner, Hrouda/66 research, the criteria for judging occupation prestige were directly based on the questions' formulation. These criteria were the occupations' social benefit, its ideal financial evaluation, assessment of the occupation (which occupation the respondents valued highest), and occupation preference in accordance with the aspirations of those questioned. The answers to the questions resulted in 4 occupation prestige rankings, which were always related to one criterion. On the basis of the occupations' assessment from these 4 criteria, it was possible to calculate an "overall social prestige", which emerged from the transfer of the coefficients referred to a uniform points scale, from 1 to 10. This overall social prestige of occupations has been chosen as the basis for a comparison with our survey.

The criteria for judging prestige in Prestige '90 were implicit in the question which aimed to find out the respondents' respect for given professions. ("Here we will be directing our attention to the prestige of individual occupations, to how people rate individual professions, in more detail"). The criteria on which the respondent based his evaluation were ascertained by a successive question, in which 12 possible answers were offered. Each of these could be given as the 1., 2. or 3. Table 2 and figure1 show how the frequency of these criteria when the occupations' prestige was judged throughout the entire population surveyed.

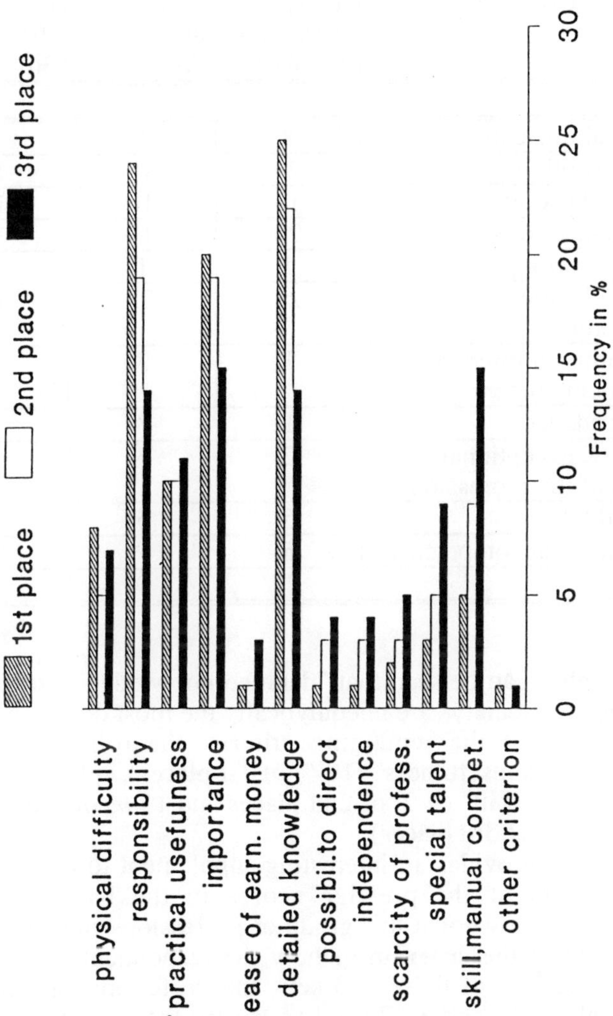

Figure 20.1 Frequency of the criteria for judging occupation prestige

Table 20.2
Frequency for the specified criteria (%) for judging occupational prestige

Criterion	Frequency of criterion specification in 1st place	Frequency of criterion specification in 2nd place	Frequency of criterion specification in 3rd place
Physical difficulty	8	5	7
Responsibility	24	19	14
Practical usefulness	10	10	11
Importance for society	20	19	15
Ease of earning money	1	1	1
Detailed knowledge requirement	25	22	14
Possibility of directing and influencing others	1	3	4
Independence	1	3	4
Scarcity, exceptional nature of the profession	2	3	5
Special talent	5	5	9
Skill, manual competence	5	9	15
other	1	0	1

From the graph it can be seen that "detailed knowledge", "responsibility" and "importance for society" are unequivocally the most dominant among all these criteria. Out of the remaining criteria, the following are worth mentioning: "practical usefulness" (10 % of 1st places), "physical difficulty" (8 % of 1st places), "skill" (9 % of 2nd places and 15 % of 3rd places), and "special talent" (9 % of 3rd places).

This information provides an interesting supplement to the results of the analysis by factors of the prestige scores in the basic group of 49 occupations. It was shown that this group of professions can be divided into two large sets when the criterion of how the respondents rate prestige is introduced. In the first of these two sets, the professions are linked with physical-manual work, and in the other the professions are, in contrast, associated with mental work. This indicates that when the respondents rated the occupations, they distinguished them according to the nature of the work, to a considerable extent (the two sets referred to represent two statistically conclusive factors for evaluation). This also corresponds to the entire occupation prestige ranking (Table 1), where the first 20 professions are predominantly of an intellectual nature, with the exception of miners, private farmers and machine operators.

The separation of these two sets of occupations from the complete group can be accounted for primarily by the fact that the criterion of detailed knowledge appears in 1st place among all the criteria for judging prestige, and then comes that of responsibility (see graph 1). As a rule, these two criteria are used in evaluating work of a mental nature.

A more detailed analysis (by factor analysis) of these two basic sets of professions showed that they can be dissected even further. The professions of scientist, professor /associate professor, research and development engineer and design engineer/engineer fall in a common subgroup of intellectual professions. This result is considerable surprising, bearing witness to the fact that public opinion intuitively grasps the nature of science's and research's creative work and its importance for society, and therefore rates these professions at a similar level.

Table 20.3
Variations in the prestige of selected occupations, 1966-1990

Prestige ranking of selected occupations Prestige '90	Occupation	Prestige ranking of selected occupations derived for comparison	Prestige ranking of occupations Brenner, Hrouda/66
1	Polyclinic doctor	1	1
2	Minister	2	8
3	Associate pro-fessor, professor (tertiary edu-cation teacher)	3	9
4	Scientist (research worker)	4	4
7	Primary school teacher (second-ary school)	5	2
8	Nurse	6	7
9	Miner	7	6
13	Lawyer	9	14
17	Design engineer, designer	8	3
18	Machine operator	10	10
22	Foreman	11	13
23	Writer, Composer	12	15

Table 20.3 (continued)

25	Bricklayer	13	18
26	Journalist	14	17
27	Tractor driver, co-operative farmer	15	5
33	Policeman	16	19
34	Lathe operator	17	12
35	Priest	18	25
37	Tailor	19	11
39	Accountant	20	16
42	Officer, Soldier	21	20
43	Secretary	22	23
45	District official	23	22
46	Unskilled worker	24	21
47	Cleaner	25	24

Table 3 shows the results of both surveys in comparison with each other. The first column lists the actual prestige rankings of 25 occupations selected for comparison from Prestige '90, and the third column gives a derived ranking. The occupations' names are taken from Prestige '90. The Brenner, Hrouda/66 survey's descriptions are placed in brackets, as the occupations here are understood slightly differently.

The correlation of both rankings listed in the table (i.e. the correlation we obtained in the occupations' ranking according to prestige, and the ranking to overall social prestige in the Brenner, Hrouda/66 survey), expressed in terms of Spearman's correlation coefficient, has a value of 0.82. This relatively low correlation coefficient shows that in the time taken, there were pronounced shifts in the ranking. However, this only concerns a few professions. The larger variations can be seen particularly in the case of ministers (a shift from 8th place to 2nd), then professors/associate professors (shift from 9th place to 3rd), lawyers (14th to 9th), priests/clerics (25th to 18th). These shifts are of a positive nature, i.e. these professions have significantly increased their level of prestige. In addition to these totally intellectual occupations, the prestige of bricklayers also increased (a shift from 18th place to 13th). The professions showing a significant shift in the opposite direction, i.e. a reduction in prestige level, were design engineers (from 3rd place to 8th) and tractor-drivers (from 5th place to 15th). In both of these examples the professions are denoted slightly differently in the compared surveys. Consequently it can be assumed that when they were rated by the respondents, a slightly different meaning was understood. In Prestige '90, the occupation of design engineer was given together with the occupation of designer, while in the Brenner, Hrouda/66 survey this was engineer-design engineer. There was a still greater difference with co-operative farmers, who are represented in Prestige '90 as tractor driver. On the other hand, the lower evaluation of co-operative farmers could be an expression of our changing economic conditions where opportunities are arising for private farming. This is confirmed to a certain extent by the fact that private farmers came 20th in the Prestige '90 survey of 49 professions.

The professions of lathe operator (a drop from 12th place to 17th) and tailor (from 11th to 19th) obtained a significantly lower rating. Some less significant variations in the occupations' placements are not without interest either. These include the increase in the prestige of writers/composers (by 3 places), journalists (likewise by 3 places), and even policemen.

The occupations of doctors (1st place), scientists (4th place), nurses (rise by one place) and miners (fall by one place) kept a high level of prestige. At the same time, those occupations which appeared in the bottom places of the survey done 20 years ago were again given very low ratings (professional soldiers, secretaries, district officials, auxiliary workers, cleaning women).

The evaluation of the occupations by their ideal earnings was an independent part of Prestige '90. Even if it can be assumed that when the respondents assessed the prestige of individual occupations they also included a financial evaluation, we were interested in whether, in spite of that, there were differences between the reputation (assessment of a profession and its ideal financial evaluation). In practice, a factually based and plausible connection between the prestige of a specific occupation and its financial evaluation is not so ambiguous. This fact can be demonstrated by the results of some surveys which have studied occupation prestige and occupation financial evaluation separately. On the other hand, the opposite approach was chosen for the surveys' ranking order, where the occupations' prestige was the result of a projection based on an investigation into ideal earnings and their comparison with actual earnings.

Only 26 professions were selected for financial evaluation from the basic group of 49, due to Prestige'90's considerable size. Respondents were asked to assign an income which they considered reasonable to individual professions from the subgroup of 26 occupations. One criterion was doing the job with current prices of goods and services staying the same. ("Imagine that you are to decide how much people can earn in specific professions. What monthly wage would you consider reasonable for the following professions? Disregard actual wage levels in the given professions, and only start from what you would consider to be fair, with the condition that the job is done with current prices of goods and services staying the same".)

Table 20.4
Evaluation of professions by ideal earnings, comparison of rankings according to amount of suggested earnings and ranking according to prestige score

Ranking according to earnings	Profession	average of suggested earnings	Ranking according to prestige score of selected professions
1	Company Director with about 2000 employees	7838	4
2	Miner working in deep mines	7583	7
3	Scientist	7070	3
4	Professor, Associate professor	6900	2
5	Polyclinic doctor	6790	1
6	Manager, Businessman	6515	8
7	R&D Engineer	5722	9
8	Private farmer	5542	12
9	Operations engineer, Technologist	5186	10
10-12	Machinist, Shop owner, Writer, Composer	4962 4930 4928	11 17 14
13	Bricklayer	4490	15
14	Foreman	4398	13
15-16	Primary teacher Car mechanic	4300 4300	5 16
17	Nurse	4141	6
18-19	Major Lathe operator	4060 4057	20 18
20	Priest	3637	19
21	District official	3138	22
22	Political party secretary	2926	25
23	Unskilled worker	2697	23

Table 20.4 (continued)

24	Secretary	2640	21
25	Cleaner	2319	24

There is a recognizable difference between the evaluation of occupations by ideal wages and their ranking according to the prestige scores obtained. The correlation of both rankings gets a Spearman coefficient of O.86. However, significant differences were only evident in the case of a few professions. E.g. doctors always come first in prestige but comes in 5th place in earnings, while miners move up from 7th place in prestige to 2nd in earnings, and company directors from 4th to 1st. We have already discussed the prestige of miners being handed down. This tradition is reflected in the financial evaluation as well, which is something that cannot be said about all the occupations. Suffice to say about the intellectual professions that their high prestige is not directly connected with a higher financial evaluation (e.g. doctors, primary school teachers). Exceptions to this are management occupations (company directors, managers) and research ones (scientists, professors, associate professors, research engineers), where the high ratings are also reflected in the financial evaluation.

The considerable difference between the prestige and earnings rankings in the case of the primary school teacher and nurse (10 grades) can be explained as follows. Primarily, both of these occupations are done by women at the moment. Wage discrimination against women is still really prevalent, as is the public's financial undervaluation of some professions requiring higher education. However, it can be said that, even with these noted exceptions, the ranking's logic in terms of afforded general respect and earnings has stayed the same, from an overall view of the ranking on the suggested wage amount and on prestige (see graph 2). Occupations coming in the first 20 on the prestige score are cited uppermost according to their earnings, and the nomination frequency for earnings over 5 000 Krs is significantly higher than for other professions.

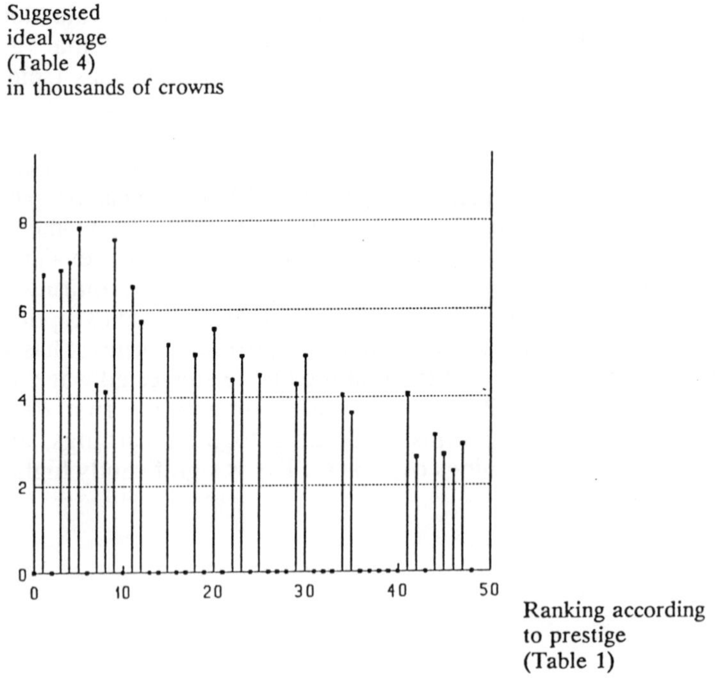

Figure 20.2 Relation between ranking of professions (according to prestige score and suggested earnings)

Table 20.5
Suggested ideal earnings and their ranking in comparable professions in the Prestige´90 and Brenner, Hrouda/66 surveys

Ranking according to earnings	Occupation	Average value of suggested earnings Prestige '90	Average value of suggested earnings B,H/66	Ranking according to ideal earnings B,H/66
1	Miner	7583	2484	3
2	Scientist	7070	2587	1
3	Prof., Associate professor, Tertiary teacher	6900	2431	5
4	Polyclinic doctor	6790	2537	2
5	R&D engineer, designer	5722	2435	4
6-7	Machinist, Writer, Composer	4962 4928	2094 1950	6 7
8	Bricklayer	4490	1837	10
9	Foreman	4398	1907	8
10	Primary/Secondary teacher	4300	1891	9
11	Nurse	4141	1382	13
12-13	Major Lathe operator	4060 4057	1658 1761	12 11
14	Priest	3637	1195	16
15	District official	3138	1419	14
16	Unskilled worker	2607	1604	13
17	Secretary	2640	1114	17
18	Cleaner	2319	977	18

A factor analysis of credit items led to parallel results as when prestige was rated by points: 3 groups of occupations emerged. In the first group are all the manual professions, joined by nurses and secretaries, in the second are the intellectual occupations, and the third group has professional soldiers, district officials and political party secretaries.

Table 5 presents a survey on whether and to what extent the public's view of a fair financial evaluation of some (comparable) professions has changed in the last decade.

There is a high correlation between the rankings of earnings in 1966 and in 1990 (0.94). It can be inferred from this that the basic earnings axis in the group of comparable occupations has been preserved. In the 24 year period under study, there were only partial shifts in suggested earnings, and then only in the case of a few professions. These differences are expressed in the rate of (discrepancy between) maximum and minimum earnings, which was 2.73 in 1966 and 3.38 in Prestige '90. This shift in public opinion reflects the emergence of new conceptions on earnings differentiation, in spite of the fact that the levelling-off of wages is continuing to date.

Conclusion

The Prestige '90 survey has provided an interesting feel for our country's social consciousness at a time of deep economic and social change.

In view of the fact that, within the framework of the selected subject, this survey was not able to cover all aspects of the points at issue, and in view of the period of time in which the investigation was carried out, it can be stated that the validity of the above results is conditional to a certain period of time. A none the less standard and methodological approach to carrying out the research, which is based on a representative selection quota of a group of respondents (1400 people), and a field investigation performed by means of a questionnaire, allow us to produce a more general interpretation and assessment of the survey's results.

First and foremost, it can be said that there is no substantiation of the introductory working hypothesis outlined at the start of the paper. While a comparison between the Prestige '90 and Brenner, Hrouda/66 surveys did establish clear shifts in occupation prestige, this was not related to research professions.

The prestige of occupations linked to scientific and research activity remains high and, specifically, a surprising discovery is professors or associate professors being placed third and scientists fourth. A high level of prestige was also obtained by the remaining occupations representing the field of research; all of these came in the top 20 occupations. This is reinforced by the results of the research professions' financial evaluation, which corresponds to their prestige rating (the other professions are not so unambiguous). This finding bears witness to the fact that these occupations are still considered socially important and valuable, even in the currently complicated social and economic situation.

The high rating of other intellectual occupations (conspicuous in the Prestige '90 results) is also connected with the distinct preference for the evaluating criteria "detailed knowledge", "responsibility" and "importance for society". The group of occupations related to activity of a scientific or research nature then formed a relatively independent subgroup of professions, assessed by the group of respondents in a similar way. This attests to the essential grasping by the public of the nature of creative work in science and their having a positive attitude towards it. (On top of this, there was no influence by age or education.)

The structure of values of our society is doubtlessly expressed in the choice of highly prestigious occupations. Health, education, responsibility and special talent are invariably placed highest in the scale of values; and this is in spite of the considerable negative influence of ideological deformations in the last 40 years.It appears that the educational tradition of a cultured people carries more weight.

The possibilities for science to solve social and economic problems were overestimated and misrepresented in the conditions of the central directive administration and the inoperation of more recognizable market mechanisms. Science was often officially held up as a panacea for social ailments, and its function with regard to world opinion was quite falsely interpreted. It was made responsible for failures over which it could not have any influence and for which the totalitarian regime was to blame. It was therefore in a very difficult situation. The respondents correctly grasped this state of affairs. The reasons for economic failures did not lie with science and research. Science is evidently perceived more as one of the victims of central directive planning and administration (shown to be lacking in vitality) and at the same time, of ideological manipulation (which has already been exposed).

Our research was not orientated in such a way as to provide sufficient data for an unequivocal answer to the question of how much some of the trends in the changes in value structures and attitudes of the population in scientifically and technically advanced countries are reflected in our modern society. These put a definite new angle on science and its social function, and there was a more demanding appraisal of science's consequences for man's quality of life, social relations and environment.

By reviewing scientific research in advanced countries, the illusion about contemporary science's "messianic role" and its "all-embracing nature" in world opinion has been removed. Instead of this, the socially relevant role of science naturally asserts itself, and is commensurate with the amount of current scientific research. Competitiveness in science is fostered and corresponds to the new expansion in market economy; in no way is this limited to a national basis, however, but is primarily on trans-national terms, integrating whole world areas. Research is becoming more and more costly, and its social and ethical consequences are often contradictory because in advanced countries a greater and greater emphasis is being placed on research activity being constantly under review, being assessed and regulated. Modern science has accepted this pressure for permanent assessment, because this increases its ability to compete in the field of research and also in a wider socio-economic context. This is the case with us as well, with analogous changes. The profundity and extent of these do not concern us in this piece of research.

The positive results of our research can also be interpreted as an expression of the expectations and wishes of our public, that in the future, science and research contribute to solving difficult problems of our times. This standpoint is an answer to the view of representatives of advanced countries that the success of a given country's economy (which is in competition with that of other countries) comes from a strong research background. In this regard, it is necessary for the public's positive attitude towards science and research to be based on a deeper and deeper awareness of their real role in society, and their possibilities and limitations in the context of European and world development.

References

Bláha, I.A. (1968), *Sociology*, Prague.

Brenner, V., Hrouda, M., 'Science and tertiary education in occupation prestige', in *Sociological journal*, 1967, no. 5 and 1968, no.1.

Hrouda, M. (1970), 'Science in occupation prestige in the Czechoslovak Socialist Republic', in *Theory and method*, no.3.

International Standard Classification of Occupations. Geneva, International Labour Office, 1975.

Machonin, P. et al. (1969), *Czechoslovak society. A sociological analysis of social stratification*. Bratislava, Epocha.

Mateju, P., Boguszak, M. (1989), *International Standard Classification of Occupations(ISCO) - Czechoslovak variation*. ÚFS CSAV, Prague.

Müller, K., Provazník, S., 'Qualitative evaluation of science, experiences of advanced countries and Czechoslovakia', in *Ideational source of scientific and techno-scientific policy in Czechoslovakia*. ÚTHV CSAV, 1990, no. 5.

Tondl, L., 'Principles of scientific and techno-scientific policy in a small country'., in *Theory of science*, 1990, no. 1.

Treiman, D. J. (1977), *Occupational Prestige in Comparative Perspective*, Academic Press, N.Y.

Afterword

A survey into the prestige of scientific (research) professions in Czechoslovak society was carried out in November 1990 as part and parcel of a research project entitled "Social Differences and the Prestige of Professions" (Prestige'90). This was conducted by the Institute of the Theory and History of Science of Czechoslovakia's Academy of Sciences on a representative sample of 1.400 respondents in the Czech and Slovak

Republics.

The results of the survey demonstrate a high level of prestige of scientific-research professions. All the five representatives of the research community finished among the top 20 professions (a total of 49 different professions were evaluated). Out of these, associate professors (professors) appeared in 3rd place in the scale of prestige, with scientists fourth and head of a research centre tenth.

As regards the prestige of scientists, it is symptomatic that, statistically, this does not depend either on the age or education of respondents. It is therefore possible to conclude that this particular profession is respected by almost all social strata of the populace.

Other intellectual professions also received high marks. This is associated with the evident preference of evaluating aspects of the represented criteria, involving "considerable knowledge", "responsibility" and "significance for society". Following on from there, the group of professions connected with activities of a scientific or research nature then forms a relatively independent subgroup of professions, assessed by the set of respondents in a similar manner. This implies that the general public essentially understands the specific nature of creative work in science, and adopts a positive approach towards it.

The results of the "Prestige '90" survey were then compared with the outcome of a research project into the prestige of professions (hereafter just Brenner, Hrouda / 66), carried out in the 1960s, which represented Czechoslovakia in an international comparison of surveys of professional prestige in 60 countries of the world.

A comparison of the above mentioned surveys has provided some other findings which are worthy of note.

A correlation (Spearman) of both professions' ranking orders according to prestige comes to 0.82, which expresses the fact that marked shifts have occurred in the ranking order in the given period. These shifts have, however, affected only some professions. Prestige increased most significantly in the profession of government minister, associate professor (professor), lawyer and priest. Significant drops in prestige were registered in the following professions: designer, tractor driver and lathe operator. A high degree of prestige was retained by the profession of physician (first place), and scientist (fourth) where there were no changes, followed by medical nurse and miner. The professions which finished in the bottom places in the Brenner, Hrouda / 66 survey once again received very low ratings (professional soldiers, secretaries, district officials, laborers, cleaning women).

The aspect of financial remuneration was studied as a separate item in the "Prestige '90 survey. Compared with the placing according to prestige, physicians finished fifth with his proposed ideal salary, but on the other hand the manager of an industrial plant advanced to first place. Both scientists and associate professors (professors) were financially assessed in a manner commensurate with their prestige position.

21 Computer diffusion. The case of university students of the University of Vienna

Ruppert W. Gaspar

1 The world of work

The deep restructuring processes which have transformed (post-)modern industrial societies in latter years come under the concept of progressive rationalization. In comparison with previous rationalization shifts, they are in turn increasingly mediated in technical terms, that is by the use of recent information and communication technologies. Economic factors have been primarily decisive in this development, such as sustained conjunctural crises and slumps, or the internationalization of markets, all of which have contributed to modify global market conditions, particularly those of capital mobilization and accumulation, inducing chains of transformation in the field of production. The new production, processing, communication and information technologies, most of which are supported by computers, are at present initiating an additional and qualitatively new series of automation. This process indeed penetrates the various economic sectors with different intensity and pace but on the whole it is leading to a general alteration of the importance given to manpower and is tending to its displacement. The expansion of the tertiary sector and the "tertiarization" of the field of production, i.e. both inter- and intrasectorial transformations of structures, are accompanied by rationalization models, themselves supported by technically modernized means of work and reduced personnel. The more so, since the continuous expansion of the fields of services and administration and the need to rationalize would prove unfeasible without the use of (personal) computers, or other computerized information and communication technologies. This means, however, that an increasing number of jobs have to deal with computers and their performance requires a "certain" qualification (now primarily in the field of software) [1].

2 The world of everyday life

Along with diverse technicalization processes operating in the professional world, an enormous multiplication of technical appliances is also observable in everyday life, outside professional activities. In this case, it is mostly a question of household techniques that make domestic chores more rational and effective, along with recreational electronics such as those used to record and reproduce visual and acoustic products. In addition, more extensively diffused information and communication technologies are being transferred from the professional to, and used in, non-professional fields of action. This is certainly conditioned, at least in part, by the temporal and spatial flexibilization processes discernible at work, which are increasingly mellowing the borderlines between work and everyday life.

No consensus has been reached in the current sociology of technology concerning the substantial determinants in the category of everyday life, which means that completely different perspectives come into the center of analysis. Still, it can be assumed for the moment that everyday life is a complementary concept to the working world and its largely "zweckrational" and strategically instrumental formative principles. Furthermore, everyday life refers to the subjective organization of life outside work.[2] In this everyday world, equipment such as telephone-answering devices, telefacsimile, mobile phones etc., have recently spread, together with personal computers. Austria has been no exception to this technicalization wave flowing into everyday life.

For several years, the process has certainly been partly accelerated in this country by the integration of computer technologies and data-processing courses in almost all branches of the educational system. Due to the fact that formal education of children has acquired increasing significance on the family scale of values[3], parents have shown more readiness to equip their school-age children with a PC to complement their learning equipment.[4] As a consequence, the distribution of so-called home computers (the "classical" device for children) as well as PCs in private households has been progressing steadily. Complementary to the increasingly rapid diffusion of such instruments in the working world, a certain compulsion has resulted forcing the individual to come to terms with these technologies in everyday life. Not only has one's attention been brought to the simple fact that these "things" are present everywhere, it has also become an everyday experience to consider PCs as a specific symbol of modernity, not only because of the publicity connected with them [5] but also as a symbol of progress, of efficient performance etc., in other (advertising) contexts. This is how more and more services and administration areas convey to their customers the idea that "absolutely nothing goes" without computers. Likewise, when the apparatus do in fact break down, the corresponding performance cannot be achieved or only so after long delays in time. To the observer, however, the relation between cause and effect in such connections is hardly evident. The work and everyday worlds alike are "soaked up" with this technology in individual and everyday worlds alike are "soaked up" with this technology in individual experiences; in turn, the fact of its being present everywhere reinforces and accelerates the diffusion process. This brings about a favorable climate which, put aside the practical advantages promised by publicity, is also strongly promoted by the media, since regular reports concerning computers and connected topics have long become obligatory in all media.

3 Between the working world and everyday life: the students

Not only schools but also institutions of higher education such as universities, technical and art schools etc., are increasingly applying PCs, even beyond those specializations and/or study orientations that would "naturally" require the use of such equipment (computer sciences, technical- and natural-science specializations etc.). Administration is still chiefly concerned in this connection but, in addition, computers are also being used to store and process specialized data in a narrower sense; this is why almost all study programs offer practical PC courses as part of their curriculum, albeit not as a requisite course. After their schooling, a large part of the students will need to decide whether they wish to undergo further training within a specific study program, especially in view of a future occupation.[6] Thus, university studies are perceived as connected in anticipation with a future occupation. Individual experiences in the world of employment include the increasing presence of computers. It is therefore understandable that a growing number of students, even if their specialization does not directly require such equipment, are getting to deal with PCs.

In 1991, 22% of all students at the University of Vienna disposed of their own computer, with a noticeable difference in sex: some 14% of the female and approx. 31% of the male students.[7] In comparison with the diffusion ratio (8%) in all Austrian private households at the time, the students represented much more than the average. There were also marked dissimilarities between various curricula: approx. 80% of the students in computer sciences, approx. 40% each in mathematics and physics, and 8%, the Austrian average, in medicine and publicist sciences. Apart from those who owned a computer, another equally large proportion (20%) had access to a computer, i.e. they could use one at some friends' or in their family. One third of all interviewees (with little difference here between male and female), who had no computer and/or limited access, were seriously thinking of purchasing one in the near future. The others, representing almost half of all subjects, did not have such intentions. In this respect, interestingly enough, age played a part contrary to the individual study orientation: the younger the subject was, the lesser intention he/she had to acquire a PC whereas the reason "lack of interest" was different from other reasons (for example, much too expensive, too little use for the studies/later profession). It is not certain, therefore, what weight those 22% may bear who, in spite of dropping costs of PCs in the last few years, have invested comparatively much money in this technology, ATS 20.000,- being estimated as an average purchase price. These costs were very high considering the financial situation of Austrian students who only dispose of an average of ATS 5000,- per month. Furthermore, 85% had a printer, which represents additional costs. Some 25% of all subjects had a part-time job and 6% were fully employed, which meant more financial security, but these two groups did not have their own computer in any larger proportions than their unemployed colleagues. For another 20%, there was no financial strain since they had received a PC as a gift. The PCs had been in possession of the subjects for an average of two years and for the large majority of them it was a first computer. In a further question concerning what had prompted the purchase of this or that particular PC, a factor analysis presented, out of nine possible arguments, two clearly differentiated patterns:[8]

1 "Rational cost-profit calculation": the individual decision had been taken according to a favorable price-performance relation.

2 "Outside influence": the influential example of friends who owned a computer or had already gained some experience with it.

To acquire a computer implies certain expectancies connected with the machine. Altogether male and female students were satisfied with their decision, their expectancies had been fulfilled and even exceeded. Only every fifteenth student declared that he/she should not have bought a computer at all because of the difficulties encountered so far; all others professed to "brand loyalty" but some wanted to acquire more efficient equipment. In spite of experienced high satisfaction, the time spent in using the computer diminishes when one owns it: the beginner's curiosity is satisfied and the computer becomes a "normal" everyday appliance.

Questions were also raised concerning utilizations, intensity, possible advantages etc., to those students who did not possess their own computer but had access to one at any time (one third of all employed students already had a computer at work!). For this reason, the database collected on such important questions as to what the computer was used for increased by another 20% (total of 678 students). After establishing the degree of usage intensity (daily, several times a week, rarely), it was found that only every eighth computer was actually used daily, almost 40% several times per week, the others more seldom. On average, males were sitting in front of their PC more frequently, whereas the age variable played an important role: the longer a subject had been involved in university studies, the more intensive the use of the computer. This is surely attributable to the increased effort invested as studies progress. In order to analyze what the machine was really used for, not only was the purpose investigated (studies, leisure time or professional occupation) but also the type of software put to work. A list of popular applications was thus drawn such as word-processing, mathematic and statistic programming, programmed languages, table calculations, databanks etc., and the degree of intensity for each usage was calculated. Only one third of all subjects also used their PC at leisure time and, if so, much more frequently than otherwise with drawing and music programs or with games (declared quota of variance with eta2 = .24). Preference for specific programs (or program packages) and intensity of usage changed if the computer was being used for activities that stood in connection with studies. At the opposite, the work activities (full or part-time employment) that forced one to sit in front of a computer did not play an important role as motive.

4 The computer as a typewriter

In this connection, a totally surprising and unexpected result was obtained: irrespective of certain differences, e.g., in connection with the field and/or duration of studies, sex etc., word-processing was the application which all computer owners (including those who had access to a machine) used almost exclusively. All other applications only played a marginal part, at least in a statistical sense. In other words, computers were concretely used as nothing more than a high-tech typewriter.

Indeed, word-processing is made much more simple with a computer as compared even to any efficient electronic typewriter. Such corrective articles as fluid, scissors, glue and others which were formerly necessary to improve a typewritten copy and made the task unpleasant are no longer needed. The price-performance relationship between computers and typewriters becomes disadvantageous, however, when computers are used almost exclusively as typewriters, even with very low-priced PCs. From a technological point of view, of course, such PC performance is only a "peripheral" one compared to what a computer "can do". In this sense, Weingart's suspicion as to the trivialization of technology can be considered: i.e. that there could be some technologies that may "become de-professionalized by ways of commercial distribution". This would happen mostly "when they are made less expensive, smaller and more simple to operate" which means "that technologies containing a potential for everyday use will undergo a clearly foreseeable development from professionalized to everyday usage, together with trivialization" (("daß Technologien, die das Potential von Alltagstechniken besitzen ...")) (Weingart 1988, p. 153).

This raises the question as to what reasons prompt students to engage in such considerable investments, i.e. how relevant are such criteria, in this sense, as calculations of costs and profit.

What is the role played by this device in a student's specific course of life; how are such equipments perceived and what other functions do they possibly accomplish beyond that of typewriters? Only one small group of subjects admitted using their computer rather often and regularly to play games and also reported having fun with this technology and being fascinated by it in one way or another. Yet this playful aspect was mostly emphasized by subjects still involved in computer-science studies.

In the opinion of 83% of the subjects, irrespective of sex, age and study orientation, the computer makes work easier and two thirds of them considered that they possessed an additional skill because they could handle a PC. Moreover, more than half of them believed that it represented an advantage in their studies and a preparation for their later professional occupation. Therefore, the PC represented a valuable working instrument the advantages of which were noticeable not only immediately but also showed relevance for one's future. Beyond this argumentative connection, and in view of the subjectively rational, however objectively rather less rational choice, the question can be raised as to what other factors could have been decisive for one fourth of the students who underwent such a high financial investment to acquire a "typewriter".

5 Many reasons - and none of them bad

At first, it seems near at hand that we are dealing with a process of rationalization according to Freud, i.e. in which a rational explanation is subsequently attributed, as logical motivation, to an action which initially appears to be completely convincing.

The facilitation of working conditions, additional skills, or the anticipation of future professional requirements could definitely be regarded as such rational constructions. But the question still remains whether a more or less competent approach to word-processing actually belongs to those future requirements, even though the drafting of texts appertains to the sphere of competence of university graduates of the most different fields of study. How this is undertaken, however, is certainly of little importance. Thus, in the case of "one-sided" usage of computers, the maxim of everyday practice "If it's no use, it won't harm" seems appropriate. Since no direct data supporting the notion of such a rationalization process are available, due to the research design, we must continue our search for further possible reasons.

The argument of facilitated working conditions is certainly justified. The preparation, correction and alteration of texts is accomplished without much effort. The modes of preparing a text can also be changed through the medium of word-processing, e.g., "recycling" texts or parts of texts already on hand poses no problem because of the existing variety of storage possibilities. New compilations, "new" texts, can easily be produced from "old supplies".[9] In addition, some kind of "bribery tactics" may also play a role, given the good readability and single reproducibility of a text. Hand-written exercises and essays are not very much appreciated by teachers due to the high amount of time lost in deciphering such manuscripts, which may even evoke an unfavorable bias towards its contents. A paper written with a typewriter certainly scores higher, especially since glued and patched-up work is easily and inexpensively improved by photocopying techniques. In comparison with typewritten and handwritten texts, however, PC-produced texts display a noticeably improved orthography, as practically all current processing systems offer so-called check-spelling functions which can compensate for individual shortcomings.[10] In the case of more extensive texts, an economic factor gains relevance which, e.g., Hennen (1992) refers to following Hans Linde. Acquiring a PC, then, is seen as an "alternative to resorting to functionally equivalent services", i.e. the costs which would otherwise be encountered for office work become economizable. The extent of such saving is not to be examined here, yet the relation between expenses and profit seems very doubtful. As noted above, the "exterior pressure" to occupy oneself with PC technology is also becoming greater in the case of students. Not only have innumerable computer shops been opened in the past few years, also in the vicinity of the University, but even an organization such as the Austrian Students Association (Österreichische Hochschülerschaft) has been running a computer shop for a while offering hardware as well as software at special rates for students. At the same time, this organization gives computer companies the opportunity to display their products within the university premises several times a year. The argument "because advertisements have convinced me" may statistically have had no bearing whatsoever upon decisions to acquire a PC (cf. Note 8), but the "ubiquity" phenomenon certainly does have an effect.

This leads us to an aspect which has recently received increased attention in the sociology of technology and which has been summarized under the heading of the culturalistic approach. Technical equipment such as computers may be considered in view of their functional (i.e. facilitating working conditions) as well as their cultural and/or symbolic content. Objects which are generally integrated in the cultural modernization process accumulate contexts of meaning beyond the purely technical one and establish a "relationship of representation" (R. Chartier 1989) which may also prove decisive in diffusion processes in students' everyday life.

As Wernick (1991, p. 69) set forth in his analysis of IBM advertising, this was "a kind of cultural language game, the ad's real content not about ... computers at all. It is about the relation of this object to a cluster of symbolized values, the computer becomes an equivalent signifier for modernity and progress". Böhme and Wehner have arrived at similar results in their analysis of interviews [11] which distinguished the following symbolic qualities of the equipment in a process of assignment to their users:

- inherent dynamics of technical development: a leading position due to the acquisition of innovative goods
- mental work and promising knowledge
- modern employment
- expert status

- orderliness, cleanliness, esthetics

This social and/or cultural "coding" corresponds indeed to the long-term perspectives that students gave as reasons for their decision to buy a computer.

To a certain extent, we can thus imply that the apparatus signal more than their bare utility in subjective perception and that they have become, on the contrary, a constituent of cultural and/or educational capital. Although computers cannot yet be considered an immediate necessity to learning and teaching in most studies and have not yet been included in curricula matter, they seem to be further perceived as a non-economic investment into the future. The symbolic capital represented by this technology may be later reversed into economic capital, should the future employment situation require so (cf. P. Bourdieu 1987, 1989), insofar as it could be connected with prospects for "better" employment. (Nowadays, it is no longer the case that the quality of a job is a problem, but rather the job itself.) To be sure, none of these various explanations can be seen as sufficient in itself while data are still missing which would clearly explain a given context. Taken together, they can serve as facets of an overall picture which does not, with the present data, appear very homogeneous nor consistent. Certainly, the notion of a rationality of individual decisions cannot be accepted in unidimensional terns, as a rationality of costs and profit. However, analyses almost suggest more differentiated forms - adequate to everyday practice - of a "rationality" which is not necessarily to be interpreted instrumentally or functionally in the sense of the world of labor. As indicated by the example of students, the significance of the cultural environment along with its technically mediated modernization aspects must be integrated into any approach to issues of technological diffusion. In the present state of knowledge, however, this implies that a great deal of explorative research must still be conducted to enable us to respond to the evident discrepancy between determinable aspects of action and individual perceptions of this action in connection with technical objects.

Notes

1. c.f. Walter Blumberger, Christina Heilbrunner, Dietmar Nemeth, Walter Lichtenauer (1990), *Chip und Job: Neue Technologien in der Arbeitswelt*, Forschungsberichte aus Sozial- und Arbeitsmarktpolitik, B.M.f.A.u.S., Hrsg., Vienna
2. c.f. esp. Bernward Joerges, Hrsg., (1988), *Technik im Alltag*, Frankfurt/Main, as well as Leonhard Hennen, *Technisierung des Alltags*, Opladen, 1992
3. c.f. Österreichischer Familienbericht 1988, Vienna

4 c.f. W. Gáspár-Ruppert (1991), *Die Spaßmaschine. Der Computer im Alltag österreichischer Schülerinnen und Schüler*, project report, Vienna

5. c.f. also Andrew Wernick (1991), *Promotional Culture Advertising, Ideology and Symbolic Expression*, London Newbury Park - New Delhi

6. c.f. Helmut Guggenberger (1991), *Hochschulzugang und Studienwahl*, Klagenfurt

7. In 1991, a total of 1592 students of the University of Vienna were interviewed in writing. The numerical basis was investigated on the number of enrollments of regular students (of all seven faculties) taking into account the sex proportion and the result was 2.25%. Distinction was made in the investigation of quotas according to an expert rating (between specialized "close to" and/or "distant from" PC in study orientations). In addition, the survey covered beginning and advanced students in order to arrive at a wider sample (dispersion) on age and number of semesters reached.

8. Factor matrix (unrotated): "Why did you choose this particular PC?"(total of 13 variables; all loads over .300 discarded)

Factor 1	
efficiency	*.61330*
advantageous price	*.54892*
so-called student special	*.36273*
Factor 2	
friends/acquaintances have one	*.59058*
others have recommended it	*.50505*
advantageous price	*.41406*
advice from specialized dealers	*.40776*

9. U. Riehm, B. Wingert, K. Böhle (1989), 'Elektronisches Publizieren - oder wie der Computer die Beziehung zwischen Autor und Verlag verändert', in W. Rammert, G. Bechmann (eds.), *Technik und Gesellschaft*, Jahrbuch 5, Frankfurt/Main - New York, pp. 86 - 104

10. c.f. also Wolfgang Coy (1989), 'Après Gutenberg. Über Texte und Hypertexte', in W. Rammert, G. Bechmann (eds.), *Technik und Gesellschaft*, Jahrbuch 5, Frankfurt/Main - New York, pp. 53 - 65

11 c.f. Wolfgang Böhme, Josef Wehner (1990), 'Der symbolische Gehalt einer Technologie', in Werner Rammert (ed.), *Computerwelten - Alltagswelten*, Opladen, p. 127

References

Blumberger Walter, Christina Heilbrunner, Dietmar Nemeth, Walter Lichtenauer (1990), *Chip und Job: Neue technologien in der Arbeitswelt*,

Forschungsberichte aus Sozial- und Arbeitsmarktpolitik, B.M.f.A.u.S, Hrsg., Wien

Böhme, Wolfgang, Josef Wehner (1990), 'Der symbolische Gehalt einer Technologie', in Werner Rammert (ed.), *Computerwelten - Altagswelten*, Opladen

Bourdieu, Pierre (1979), *Entwurf einer Theorie der Praxis*, Frankfurt/Main

Bourdieu, Pierre (1987), *Die feinen Unterschiede*, Frankfurt/Main

Chartier, Roger (1989), *Die unvollendete Vergangenheit. Geschichte und Macht der Weltanschauung*, Berlin

Coy, Wolfgang (1989), 'Après Gutenberg. Über Texte und Hypertexte', in W.Rammert, G. Bechmann (eds.), *Technik und Gesellschaft*, Jahrbuch 5, Frankfurt/Main - New York

Gáspár-Ruppert, W. (1991), *Die Spaßmaschine. Der computer im Alltag österreichischer Schülerinnen und Schüler*, Projektbericht, Wien

Guggenberger, Helmut (1991), *Hochschulzugang und Studienwahl*, Klagenfurt

Hennen, Leonhard (1992), *Technisierung des Alltags*, Opladen

Joerges, Bernward, Hrsg.,(1988), *Technik im Alltag*, Frankfurt/Main

Österr. Familienbericht 1988, Wien

Riehm, U., B. Wingert, K. Böhle (1989), 'Elektronisches Publizieren - oder wie der Computer die Bezeihung zwischen Autor und Verlag verändert', in W.Rammert, G. Bechmann (eds.), *Technik und Gesellschaft*, Jahrbuch 5, Frankfurt/Main - New York, S.86 - 104

Wernick, Andrew (1991), *Promotional Culture - Advertising, Ideology and Symbolic Expression*, London - Newbury Park - New Delhi